LIVING *in* INFORMATION

Responsible Design
for Digital Places

T0272959

by

JORGE ARANGO

foreword by Hugh Dubberly

TWO WAVES
BOOKS

TWO WAVES BOOKS
BROOKLYN, NY, USA

Living in Information
Responsible Design for Digital Places
By Jorge Arango

Two Waves Books
an Imprint of Rosenfeld Media, LLC
540 President Street
Brooklyn, New York
11215 USA

On the Web: twowavesbooks.com
Please send errors to: errata@twowavesbooks.com

Publisher: Louis Rosenfeld
Managing Editor: Marta Justak
Interior Layout Tech: Danielle Foster
Cover and Interior Design: The Heads of State
Indexer: Marilyn Augst
Proofreader: Sue Boshers

© 2018 Jorge Arango
All Rights Reserved
ISBN: 1-933820-65-9
ISBN-13: 978-1-933820-65-1
LCCN: 2017958794

Printed and bound in the United States of America

For Boisie,
who introduced me to information environments.
I wish he could have experienced this one.

Contents at a Glance

Foreword by Hugh Dubberly   x

Introduction   xiv

Chapter 1. Environments   1

Chapter 2. Context   21

Chapter 3. Incentives   33

Chapter 4. Engagement   47

Chapter 5. Technology   65

Chapter 6. Architecture   81

Chapter 7. Structure   99

Chapter 8. Systems   119

Chapter 9. Sustainability   135

Chapter 10. Gardening   153

Conclusion   169

Index   175

Acknowledgments   185

About the Author   187

Colophon   188

Contents and Executive Summary

Foreword by Hugh Dubberly   x

Introduction   xiv
We are in the midst of a major social transformation—moving many of our day-to-day activities from physical places to information-based places that we experience on our phones and computers. The central question of this book is: *How can we design these information environments so they serve our social needs in the long term?*

Chapter 1. Environments   1
The form and structure of our environments shape our interactions with each other and with our social institutions. For most of our history, we've operated within physical environments. But now we are also inhabiting environments that are made of information.

Chapter 2. Context   21
Our relationship with our environments establishes contexts that influence our thinking and behavior. Our awareness of where we are and what we can do there is informed by affordances and signifiers in the environment.

Chapter 3. Incentives   33

The form and structure of our environments have not emerged arbitrarily; instead, they have developed over time to help us fulfill particular needs. These needs are driven by incentives that influence both design constraints and intended user behaviors.

Chapter 4. Engagement   47

The business model that drives today's most popular information environments incentivizes users to pay attention to the environment itself instead of each other. This leads to social dysfunction.

Chapter 5. Technology   65

In addition to business incentives, technology also influences the form and structure of our environments. Virtual and augmented reality, artificial intelligence, voice-based user interfaces, and the blockchain are five current technologies that promise to change the structure of our environments and how we experience them.

Chapter 6. Architecture   81

We can intentionally design our environments to better serve our needs. Architecture is the design discipline that is focused on structuring our physical environments, and information architecture is the design discipline that does the same for information environments.

Chapter 7. Structure   99

Architects define the conceptual structure of our environments, which is perhaps the single most important factor in how we experience them. In information environments, these structures manifest as labeling and navigation systems that impose distinctions between parts of the environment.

Chapter 8. Systems   119

Environments are not just structural constructs; many other systems must work in concert to make it possible for them to serve our needs. Architects must consider how these systems work together.

Chapter 9. Sustainability   135

These systems are constantly changing. Architects must vie to make them evolve in ways that don't compromise their integrity or usefulness in the long-term. Some parts of the environment evolve more slowly than others; long-lasting environments establish structural distinctions that generate social, economic, and ecological value.

Chapter 10. Gardening   153

Ultimately, creating environments that support our needs in the long term requires that we relinquish top-down control in favor of a more generative approach. Such an approach encourages emergence and continual evolution while preserving integrity and generating value.

Conclusion   169

Index   175

Acknowledgments   185

About the Author   187

Colophon   188

Foreword

Designing Within Systems

Designing has its roots in craft—in making "things," in giving them form. And at one level, designing is concerned with "how things look"—their shape, color, and material. Yet, while "good form" is important, form is not the only concern in designing. Designers are coming to realize that "things" are enmeshed in networks—"gathered together" in systems—biological systems, systems of goods and trade, information systems, social systems, systems of technology, and more. And increasingly, designers are recognizing that we are designing within systems.

Recognizing that design has several dimensions has a long history.

Roman architect Vitruvius described three principles: "durability, convenience, and beauty." The International Standards Organization (ISO) echoes Vitruvius, mandating software that is "effective, efficient, and engaging." Architect Louis Sullivan proclaimed, "form ever follows function"—while Frog founder and Apple product designer Hartmut Esslinger quipped, "form follows emotion."

Apple co-founder Steve Jobs, who hired Frog early, noted, "In most people's vocabularies, design means veneer. It's interior decorating. It's the fabric of the curtains and the sofa. But to me, nothing could be further from the meaning of design. Design is the fundamental soul of a man-made creation that ends up expressing itself in successive outer layers of the product or service."

Jay Doblin, co-founder (with Massimo Vignelli) of Unimark International, one of the first corporate identity firms, described the form and function dimensions in terms of appearance and performance. Doblin proposed a 2 x 3 matrix of "six types of design" with appearance and performance on the Y axis and products, unisystems, and multisystems on the X axis. Doblin describes "products" as "tangible objects" and "messages"; unisystems as "sets of coordinated products and the people who operate them"; and multisystems as "competing unisystems."

Richard Buchanan, who has a PhD in rhetoric from the University of Chicago, for many years headed CMU's design school, and now teaches in the business school at Case-Western, has proposed a similar framework of four "spaces" or "orders" of design: communications (a focus on meaning and symbols); artifacts (a focus on form and things); interactions (a focus on behavior and action); and fourth order (a focus on "environments and systems in which all other orders exist").

Michael Porter, who teaches at Harvard Business School, has written about "how smart, connected products are transforming competition" and "redefining industry boundaries." Porter described a similar framework with five-phases: 1) products, which become 2) smart products, which become 3) smart-connected products, which join 4) product systems, which join 5) systems of systems. Increasingly, value comes from adding "intelligence" to products—microprocessors, software, and sensors. Further value comes from connecting products to cloud-based processing, networked applications, and human services—what CMU HCII head Jodi Forlizzi calls *product-service systems* or *product-service ecologies*. Examples might be Apple's iPhone–iTunes–App Store ecology or similar ecologies offered by Amazon, Facebook, Google, Microsoft, and others.

John Maeda, former President of Rhode Island School of Design (RISD), has offered a sort of era analysis, suggesting design practice has evolved in three stages: 1) classic design ("perfect, crafted, and complete"), 2) design thinking ("innovation…experience …empathy"), 3) computational design ("design for billions of individual people and in real time is at scale and TBD"). Design thinking clearly has roots in systems thinking, as does computational design (e.g., artificial intelligence, machine learning, deep learning, natural language processing, computer vision, etc.).

Joi Ito, head of MIT's Media Lab, has summed up the shift in what, how, and where we design. "Design has also evolved from the design of objects both physical and immaterial, to the design of systems, to the design of complex adaptive systems. This evolution is shifting the role of designers; they are no longer the central planner, but rather participants within the systems they exist in. This is a fundamental shift—one that requires a new set of values."

The models proposed by Doblin, Buchanan, Porter, Maeda, and Ito each provides a lens on an ongoing shift in design practice. While the frames are by no means analogs, they each in their own way point to an expansion of design practice from a narrow focus on things to a broader view of systems. And several of them recognize explicitly that designers are enmeshed within those systems.

Recognizing that we are designing within systems is not new. In 1969, Gordon Pask wrote, "… a building cannot be viewed simply in isolation. It is only meaningful as a human environment. It perpetually interacts with its inhabitants, on the one hand serving them and on the other hand controlling their behavior. In other words structures make sense as parts of larger systems that include human components and the architect is primarily concerned with these larger systems; they (not just the bricks and mortar part) are what architects design."

While the idea that we are designing within systems is not new, most designers are just now discovering its truth and relevance. We are just now beginning to grapple with how to design within systems. And that means we need not only new values (as Ito suggested) but also new tools.

Jorge Arango has given us both. In the book that follows, Arango offers an introduction to designing within systems. He argues that we are "living in information"—in virtual structures that serve us *and* control our behavior. Drawing from the principles of physical architecture, he suggests principles for virtual architectures. He points out that "things" both respond to "context" *and* shape it. And he reminds us that we are responsible for our language—and for both the "things" that we design *and* their "contexts." He asks us to take the long view—a whole-systems view.

—Hugh Dubberly
Principal, Dubberly Design Office

Introduction

Every weekday morning I commute to work on BART, the San Francisco Bay Area's metro system. When I look around at my fellow passengers, I'm struck by how few of them seem to be fully present. Many of them—often, most of them—are staring into small glass rectangles in their hands. Their bodies share this train car with me, but their minds and attention are elsewhere.

Where are these people? I sometimes catch glimpses: chat bubbles, games with colorful candy explosions, videos of Bollywood dancers, a news website, a cat GIF, Facebook. Sometimes a smile or a frown flashes across their faces, evidence of an interaction that the rest of us are not privy to. Their focus is intense: they only come back to the here and now when the train pulls into a station or makes an unexpected stop.

In the moments during which these passengers are focused on their glass rectangles, we've somehow stopped being in the same place together. The boundaries of the physical environment we share no longer constrain them: they're engaged in something—a bank transfer, a political argument, a shopping expedition, a flirtatious encounter—that's happening somewhere else. That somewhere is very interesting to me. This is the "place" where many of us do our shopping, learning, and banking. We meet with our friends and loved ones there. It's also where major parts of our social and civic interactions are playing out. Every year we're spending more of our time there. According to a survey by We Are Social and Hootsuite, as of 2017, people spend an average of five hours and twenty minutes online every day, with some countries reporting an average of nine hours per day. That's over half of their waking hours spent in glass rectangles of various shapes and sizes.[1]

1. https://wearesocial.com/special-reports/digital-in-2017-global-overview

Since 2016, over 4,000 retail stores have closed in the U.S. The trend, which has been called "the great retail apocalypse" in the media, is being driven in part by a shift in shopping habits from physical stores to websites such as Amazon.com.[2]

It's not just retail. More of us are also finding our mates online. According to a study published by the Pew Research Center, the share of 18- to 24-year-olds in the U.S. who used online dating sites tripled between 2013 and 2016. Today, 5% of Americans who are married or in a committed relationship say they met their significant other online.[3]

Americans are also increasingly learning online. According to a study published by Babson College and sponsored by several other institutions, more than six million students registered for online courses in 2015. This represents almost 30% of all higher education enrollments in that year. The percentage of students signing up for such "distance education" courses has been increasing, while on-campus student registrations have been declining.[4]

And, of course, we're also having more of our public discourse online. The influence of social networks such as Facebook and Twitter in the 2016 U.S. presidential election has been well documented.

2. "What in the World Is Causing the Retail Meltdown of 2017?" *The Atlantic*, April 2017, https://www.theatlantic.com/business/archive/2017/04/retail-meltdown-of-2017/522384/
3. "5 Facts About Online Dating," Pew Research Center, February 2016, http://www.pewresearch.org/fact-tank/2016/02/29/5-facts-about-online-dating/
4. "On Campus Enrollment Shrinks While Online Continues Its Ascent." *Campus Technology*, May 2017, https://campustechnology.com/articles/2017/05/02/on-campus-enrollment-shrinks-while-online-continues-its-ascent.aspx

We're moving more key social interactions to information systems every day. Although I speak of "small glass rectangles," this is just a synecdoche for all the digital devices that give us access to these environments. A decade or so ago, the majority of people accessed them using desktop or notebook computers; today, most people use smartphones. In the future, we may access them using devices with no glass at all, such as smart speakers or smart objects embedded in our physical surroundings. Technology proceeds unabated, weaving software into more and more parts of our daily lives.

Given how central these software-based experiences have become to our societies, we must ensure that they serve human (and, more broadly, planetary) needs. We must look beyond the alluring superficial aspects of what technology can do for us, to the underlying contexts and systems we're creating and to the distinctions we're imposing on the world. We must strive to make these systems viable in the long term and ensure that they also support the viability of the societies that make them possible.

Most contemporary discussion about software design frames the object of the work as a product, a tool, an interaction, or (at best) a service— all transactional and, to a greater or lesser degree, ephemeral. Software applications do have characteristics of all of these things, but they also have characteristics that make them place-like; they create contexts that influence the way we understand the world and, hence, how we act in it.

Places are longer-lived than products, services, or tools; we conceive of things differently when we know they must endure. We've also been designing places—buildings, towns, parks, etc.— for a long time. We understand the forces that shape spaces and forms, and how they influence our behavior. As a result, there's much that software designers can learn from architecture.

This book aims to make these connections explicit. In particular, it seeks to answer the following question: *How can we design information environments that serve our social needs in the long term?*

This calls for software that fosters sustainability and resilience at all levels: economically, socially, and ecologically.

In the *Spider-Man* stories, Peter Parker learned that with great power came great responsibility. Software designers are facing a Peter Parker moment: we must realize the great power we have over people's understanding of the world and their behavior in it, and wield that power responsibly. We must—in Alan Cooper's memorable phrase—become better ancestors.[5]

A Bit About Me

I've been designing software (mostly websites and apps) professionally for almost 25 years, and as a hobby for at least a decade before that. I was educated as an architect, and worked as one for a year before I left architecture to dedicate myself fully to designing information environments. So my approach to user interface design is informed by placemaking. As you'll see, there is much that software designers can learn from architecture.

Who This Book Is For

"But," you may protest, "I'm not a software designer!" Don't worry—I'm not talking about people who have the word *designer* in their job title. As you'll see, I have a rather broad understanding of what design is. If you are responsible for a digital product or service, or part of a team responsible for one, you will benefit directly from understanding how to design more sustainable information environments. And if you aren't responsible for such a system, you will still benefit from reading this book. Many of the most important decisions in your life are mediated in places that happen in small rectangular screens. It behooves you to understand how information environments affect your behavior.

---

5. Cooper has issued this challenge to designers in various online posts and keynote speeches, most recently at Interaction 2018 in Lyon.

We live today not in the digital, not in the physical, but in the kind of minestrone that our mind makes of the two.

—Paola Antonelli

1

Environments

One Montgomery Street in San Francisco is home to a branch of the Wells Fargo bank. It's also a time portal. When you step through its semi-circular portico, you're transported back to 1908, the year the building opened. Designed by renowned San Francisco architect Willis Polk to be a bank, One Montgomery looks the part. Its exterior is staid sandstone punctured by a series of tall arched windows, tied together by an ornate frieze. The interior is all business—from a time when business was a less hurried, more elegant affair. While it still serves as a functional bank branch, One Montgomery is not what 21st century bank patrons expect. It's a cathedral for business, with high, ornate ceilings, soaring marble columns, low lighting, and a hushed tone. Mary Poppins's George Banks would feel right at home.

One Montgomery Street in San Francisco.

PHOTO BY BEYOND MY KEN VIA WIKIMEDIA COMMONS,
HTTPS://COMMONS.WIKIMEDIA.ORG/WIKI/FILE:2017_1_MONTGOMERY_STREET.JPG

I toured One Montgomery in the spring of 2017 with an architectural historian who pointed out the details that made a bank in the first few decades of the 20th century. A raised central station, to allow branch managers to oversee staff. Marble cladding. Carved stone tables for customers to fill out their deposit slips, featuring built-in inkwells.

The inkwells have long been dry, since most people don't write with fountain pens anymore. And more often than not, they don't "bank" in buildings such as One Montgomery. Today, most of our financial dealings—and many other activities—happen in a different type of environment, one in which we enter and leave on a whim through screens we carry around in our pockets or unfold on tables in coffee shops.

Whether they be websites on your notebook computer, apps on your phone, or "conversations" with the "smart" cylinder on your mantelpiece, these environments are where you catch up with your friends, work, study, find a romantic partner, bank, shop, and undertake a whole host of other activities that our forebears did in physical space. Because they are composed primarily of information—words and images on screens—we refer to them as *information environments*.

We know how to design and use physical banks such as One Montgomery. We've been using places for this and other purposes for thousands of years. The forms of buildings have evolved over that time to suit our needs. However, information environments are still new. Patterns for their effective use are only now starting to evolve. As with every new medium, we bring to information environments biases and expectations that are not inherent to them, but echoes of the past. Let's start by looking at how we use places: parts of our physical environment that we've set apart for particular uses.

Physical Environments

What do you understand by *environment*? If you're like most people, the word will evoke images of the rainforest, whales breaching the surface of the ocean, or smokestacks spewing filth into the atmosphere. In other words, ecological images. This is not surprising, since you often see *environment* in phrases such as "protect the environment" or "save the environment" or "environmental pollution."

The natural environment is certainly an example of what I mean by "*environment.*" However, I also mean it a bit more generally. When I say *environment*, I mean the "surroundings of a system or organism," especially the aspects of those surroundings that "influence the system's or organism's behavior." (This latter condition is important; you could say your surroundings include all of the solar system, but the orbit of Jupiter has very little influence on your day-to-day actions.[1])

We exist in a physical environment. Large parts of this environment are natural and wild, untouched by human civilization. However, most of us spend our lives in physical environments that have been reconfigured by other people toward particular ends. These "artificial" physical environments—buildings, parks, streets, towns, cities, etc.—have a

---

1. You may disagree if you're into astrology. (I would argue in that case that it is the *belief* that the orbit of Jupiter influences your actions, not the planet's orbit per se.)

great influence on our behavior. They make it possible for us to collaborate with our coworkers, sleep soundly at night, or quietly read a book.

Beyond these obvious sheltering functions, physical environments also play important psychological and cultural roles. Think of your favorite restaurant, one you return to time and again. What makes it special? Is it the taste of the food? The comfort of its chairs? The quality of service? How cheap it is? Its proximity to your home or office? The design of its architecture? Memories of good times you've had there? Perhaps it's a combination of some or all of these factors. Whatever it is, there are many other environments in the world set apart for eating and drinking, but this one is special to you.

When you inhabit such an environment, use it for its intended purpose, and interact with other people there, it becomes part of your mental model of the world. You start using it as a reference point in your own personal geography. When you are there, you feel, think, and act in ways that are particular to that environment. We call such environments *places*, and they are central in our lives.

Humans have been setting apart space for particular uses for many thousands of years. This cave art is from the Cueva de las Manos in Argentina and was painted between 13,000 to 9,000 years ago.

IMAGE: HTTPS://EN.WIKIPEDIA.ORG/WIKI/FILE: SANTACRUZ-CUEVAMANOS-P2210651B.JPG

As a civilization, we've been setting aside places for particular uses for a long time. Perhaps the first physical place was a clearing in a forest, or an opening in the side of a mountain, where a small group of people gathered to eat and rest. As our cultures evolved, we developed more sophisticated and specialized places. For example, a place of worship calls for a different setting than a bustling market does. As a result, building types and techniques evolved to meet these and other needs over time. Your restaurant is among the latest manifestation of the "eating place" that we've been building for thousands of years.

Our effectiveness as individuals and societies greatly depends on how well these places serve the roles we intend for them. You may have experienced the effect that your environment has on your performance firsthand if, like many people today, you work in an open office. A friend of mine is always complaining about having to work in such a "cube farm"; her coworkers make constant noises that destroy her concentration. The quality of her work in such an environment will be different from what she would produce in a place that allowed her greater control over her attention.

While their primary purposes may be to give us safe locations to work, eat, learn, worship, and more, many places also meet another important need as well: they allow us to come together as a community. In his book *The Great Good Place*, sociologist Ray Oldenburg coined the term "third place" to refer to places such as cafés, bookstores, hair salons, and bars that are not our home (the "first" place) or work ("second" place), and which allow us to socialize with our neighbors and fellow citizens. For example, in many small towns in the U.S., the local post office is the place where people catch up with their neighbors and get a sense of what they think and feel about the state of the world. They may come for the mail, but they also get an understanding of where their community stands on various issues of the day.[2]

---

2. Ray Oldenburg, *The Great Good Place: Cafes, Coffee Shops, Bookstores, Bars, Hair Salons, and Other Hangouts at the Heart of a Community* (Cambridge: Da Capo Press, 1999).

Some places also tell stories about who we are—and who we were—as a people. Consider the Centro Cultural Gabriela Mistral, or Centro GAM, in Santiago, Chile. This is a modern building complex that has had various uses during its existence, all of which speak to Chileans in particular ways. It was built in 1972 to serve as a convention center shortly after the election of then-new president Salvador Allende. Beyond this utilitarian purpose, it was also meant to function as propaganda for the new government—it was built in under a year by volunteers after an extraordinary effort. When Allende was overthrown, the Centro GAM became the seat of the military government led by Augusto Pinochet. Parts of it were used later to house the headquarters of Chile's defense forces. In 2006, a considerable part of the complex was destroyed in a fire; however, when I last visited Santiago in 2016, a major restoration of the complex was almost finished. As a physical structure, the Centro GAM is a container for particular activities, but for Chileans, it's also a container of the country's modern history.

The Centro GAM was still being refurbished when I visited Santiago in the fall of 2016.

So places serve on two levels: they perform physical functions and symbolic ones. Both are essential to healthy societies. Physically, they shelter us and provide us with contexts in which we can effectively perform our activities, including the secondary, but no less important, activity of socializing. Symbolically, they embody and catalyze our cultural identities at the local, national, or global level; in other words, they ground us.

On both levels, places convey information. At the physical level, a building's form conveys to your senses the possibilities for action that it makes available to you. A wall keeps meetings private. An opening on the wall allows you to cross through to the other side. A sidewalk encourages you to walk in a particular direction. A bolted door makes it impossible for you to enter (and lets you know that's the case). A glass storefront gives you a preview of the goods sold inside. Your senses take in these physical features of the place automatically; they let you know what you can and can't do there.

At the symbolic level, places convey information by using location, scale, symmetry, rhythm, material selection, and more, to establish their relationship to other elements in the environment. If you've ever visited the National Mall in Washington, D.C., you've experienced the power of an architecture designed to convey symbolic information about the place.

In the National Mall, the location of buildings with relation to open spaces, their relative sizes, the materials used in their construction, their architectural language, and so on have been carefully chosen to have a specific effect on you. These buildings' forms provide much more than mere spaces for people to debate and enact the laws of the United States. The particular effect they have on you will depend on many factors, starting with whether or not you are a U.S. citizen.[3]

---

3. For an excellent overview of how information in the environment affects our behavior, see Andrew Hinton's *Understanding Context* (Sebastopol: O'Reilly Media 2014).

The National Mall in Washington, D.C.

PHOTO BY JOHNNY BIVERA, PUBLIC DOMAIN, HTTPS://COMMONS.WIKIMEDIA.ORG/W/INDEX.PHP?CURID=32519919

To summarize, physical environments both convey information and create the contexts necessary for people to exchange information with each other. In a very real sense, buildings and cities are the original social networks. They're also cultural manifestos in stone and terra cotta; they keep relating stories long after the people who created them left the scene. Given how central placemaking has been to our species, and the degree to which places work for us regardless of our level of education, it is no exaggeration to claim that architecture was our earliest, most enduring, and perhaps most important information technology.

Chartres Cathedral conveys information about man's relationship to the divine through the configuration of space in and around the building, and through more literal carvings on its surfaces.

PHOTO © GUILLAUME PIOLLE, CC BY 3.0, HTTPS://COMMONS.WIKIMEDIA.ORG/W/INDEX.PHP?CURID=10219594

Information

We need to take a step back here. I've said that physical environments convey information and serve as contexts where people convey information to each other, and that places are an information technology. I've also talked about "information environments," and suggested they are different from physical environments. Before we go any further, we need to look more closely at this word "information" to make sure we're on the same page.

You normally think of information as something you find in books, newspapers, and websites; the stuff in the world that adds to your knowledge. You talk about living in the "Information Age" and being "information workers"; your phones and computers are "information technologies." But information is not only something you learn through books and websites, but it's also part of your surroundings. In fact, you couldn't make sense of the world without it. There's information all around you at this very moment. So what is it?

You can think of information as anything that helps reduce uncertainty so that you can make better predictions about outcomes. That's somewhat abstract, so let's look at a pedestrian example. Every morning I walk my dog, Bumpkin, around our neighborhood. Most of the houses where we live have front yards. The owners of some of those houses have placed signs on their yards that look something like this:

Information happens.

IMAGE BY DAVID SWAYZE, VIA FLICKR,
HTTPS://WWW.FLICKR.COM/PHOTOS/SWAYZE/3195122793

Whenever I encounter a yard with one of these signs on it, I know I shouldn't allow Bumpkin to poop there. It's not that he can't poop there; physically nothing bars him from going on the yard. Rather, the sign helps me predict a likely outcome of my decision to let him do it; namely, having to deal with an irate homeowner.[4] The sign provides information about that particular yard; it sets a value for an attribute of the yard that sets it apart from the others around it. (You could express it in pseudocode: PoopHere = FALSE.)

Note that this doesn't mean the owners of yards that lack these signs think it's OK to let dogs poop on them; they've taken no formal position on the matter one way or another.[5] If I were to let Bumpkin poop in one of the yards with no signs, I could face an irate homeowner—or not. The yards with "no poop" signs on them have merely reduced my uncertainty on this matter with regard to that small part of the universe. Thus, they provide information that influences my actions when walking my dog.

It's easy to see how signs provide information, but what about other aspects of the environment? You get lots of information from other parts of your surroundings that also influence your actions. For example, many of the forms around you have been designed to let you know how they are meant to be used. Consider how the entrance of most public buildings is carefully designed so that you can easily find it, even if you've never visited that particular building before.

---

4. Note that I, not Bumpkin, am the intended recipient of the information. Even though ultimately it is his behavior that is being influenced, the homeowner assumes that dogs will be leashed and under the control of their owners. The "no poop" signs would be meaningless to packs of wild dogs roaming the neighborhood.
5. You could argue that the first homeowner who put up a "no dog poop" sign imposed the duality on all the others: suddenly, front yards were divided into those that allowed pooping and those that didn't, whether they wanted this or not.

Entrance to the High Court at Chandigarh, by Le Corbusier.

PHOTO: HTTPS://WWW.FLICKR.COM/PHOTOS/INFANTICIDA/6204214446

Architects highlight the point of entry to a building by recessing openings, creating deep shadows with cantilevered roofs, breaking the rhythms of the facade, or changing the roofline, among other techniques. Even though these aren't literal signs in the same way that the "no poop" signs are, they're visual cues that tell you something is happening at that point in the structure. They help reduce your uncertainty, and hence improve your ability to act. They provide information.

Information Environments

You may be wondering: If information is present everywhere around us, why make the distinction between physical and information environments?

Over the course of our history, our species has produced technologies that have improved our ability to communicate, store, and process information. The first—and still most important—of these is language, at first spoken and eventually written. Language allowed us to inform one another over space and time. You needn't have been born in Rome around 60 BCE to benefit from Lucretius's wisdom; written language allows the information he compiled to bridge the gap between your two lifetimes.

Over time, these information technologies have become better, faster, cheaper, and more ubiquitous. Paper scrolls were an improvement over clay tablets, codices an improvement over scrolls, printed books over manuscripts, and so on. Eventually, the telegraph allowed us to tap electricity to transmit information instantaneously over long distances. This enabled people to communicate in (almost) real time while being in different physical places. The telegraph was followed by a series of ever-more-powerful information technologies: wireless radio, the telephone, and television, to name the most important.

Then, in the middle of the 20th century, a new information technology came along that would change everything: digital computers. Born of war, computers were initially conceived as super-powerful calculators to guide missiles. However, it soon became apparent that they could help us with all sorts of tasks that could be represented symbolically—even those that didn't specifically deal with numbers.

Over the last five decades of the 20th century, computers became ever smaller, cheaper, and more powerful, and were eventually connected into a vast network that amplified their usefulness and power in previously unimagined ways. This computer network—the internet—has become central to our lives. We depend on it for all sorts of tasks, from keeping in touch with our loved ones to shopping to working to finding a mate. Some of us even wear internet-connected computers on our bodies, where they track our activities and occasionally prompt us to exercise.

Consider what happens when you chat with a friend using an app such as Apple Messages in one of these internet-connected devices. You and your friend are communicating in real time, even though your bodies may be physically very far from each other. While you're chatting, neither of you are focused on your physical surroundings. Instead, your minds are operating within a context that's defined by the chat app; the two of you are represented in the space as little images within circles, your words conveyed by speech bubbles, much like cartoon characters.

The chat application becomes your shared environment, its boundaries defined by the app's user interface much as the boundaries of a physical room are defined by its walls and ceiling. You and your friend are sharing this environment, even though you're not physically in the same place. This environment is made almost entirely of information; you can't eat or sleep or exercise there. (But you can find out where you're going to eat, how deeply you've slept, and how much you've exercised.) Hence, while you're chatting, the two of you are inhabiting a shared information environment.

Physical environments are not all the same. A conversation held in a confessional in a church has a very different character than one held in a beauty shop or coffee house. The same is true of information environments; a conversation that happens in Apple Messages (where you're afforded some degree of privacy) will have a different character than one held over Twitter, which is more public. Information environments create contexts that influence our behavior and actions.

The writer and designer Edwin Schlossberg said, "The skill of writing is to create a context in which other people can think."[6] I think the skill of designing—especially designing software—is creating contexts in which other people can work, learn, play, organize, bank, shop, gossip, and find great gelato. We're in the process of moving many of these activities, which we have heretofore realized in physical environments, online. The impacts of this transition on important aspects of our lives—how we shop, work, learn, and more—are worth considering. Let's look at some now.

6. Cited by Tim O'Reilly in *Tim O'Reilly in a Nutshell: Collected Writings of the Founder of O'Reilly Media* (Sebastopol: O'Reilly Media, 2011).

Shopping in Information

Consider the phenomenon that is being called "The Great Retail Apocalypse of 2017": a massive closure of physical retail shops in the U.S. Over 4,000 locations were affected, with some retailers such as Payless Shoe Source, Sports Authority, RadioShack, The Limited, and Wet Seal declaring bankruptcy. Major players, such as JCPenney, Sears, and Macy's have closed over 100 stores each, with the latter eliminating 10,000 jobs as a result. According to an article in *The Atlantic*,[7] the simplest explanation is the rise of online retailing, particularly through Amazon.com, whose sales in the North American market quintupled from $16 billion in 2010 to $80 billion in 2016. Shopping has always been grounded in information. The buyer who has less information about prices than the seller is at a disadvantage. Information environments such as Amazon do a better job than physical shops as settings for the sort of information arbitrage that happens in a commercial transaction. When you shop for something in Amazon, you are a much better informed—and therefore, more powerful—purchaser than if you shop in a physical store. The economies of scale that come from serving a larger customer base lead to lower prices. Since the system is freed from the constraints imposed by physical stores, it can offer much more diverse inventory. And because the environment is made of information, it can reconfigure itself dynamically to make the relevant parts of this inventory more easily available to each individual customer. The combination of these factors is difficult for physical retail stores to compete with.

Working in Information

Much of our work, too, is increasingly happening in information environments. Many white-collar jobs require that people spend significant amounts of their time focused on their computers and phones,

7. https://www.theatlantic.com/business/archive/2017/04/retail-meltdown-of-2017/522384/

interacting with each other through information environments such as Slack, Outlook, and Salesforce.com. Today, many of us work with collaborators in different parts of the world, some of whom we don't get to interact with in physical space at all. Our interactions with these people are completely mediated through screens and (less frequently) speakerphones. The way we organize our shared information environments has (at least) as big an impact in our ability to collaborate as the way we organize our physical offices.

Learning in Information

Education is also moving to information environments. At a time when the cost of traditional higher education is rising,[8] MOOCs (Massive Open Online Courses) and online learning providers such as Coursera, Udemy, and Khan Academy offer a lower-cost alternative. Major universities such as Stanford, Harvard, and MIT already offer such courses. But it's not just higher education that is undergoing this transformation. My grade-school daughters cheerfully talk about taking assessment tests in their Chromebooks at school. And much corporate training happens in learning management systems that allow employees to learn at their own pace and their managers to track their progress.

Socializing in Information

Increasingly, we socialize and get a sense for what's going on by interacting in information environments. Social networks such as Facebook (which as of December 2017 had 2.17 billion active monthly users), WhatsApp, Snapchat, and Twitter are where many of us catch up with our friends today. Whereas Oldenburg's third places reinforce a sense of local community by being grounded in a particular place,

---

8. https://www.washingtonpost.com/news/grade-point/wp/2016/10/26/college-costs-rising-faster-than-financial-aid-report-says/?utm_term=.e827e1c045e0

digital social networks are unfettered by such constraints.[9] The opinions we're exposed to in these systems are not those of our neighbors, but those of the people whom algorithms have determined will keep us engaged. The effects on democracy of having citizens inform their world views in such environments is a topic of ongoing study. That said, I can confidently say that engaging with each other in a context where over a quarter of the world's population is present is bound to have some effect on our ability to act collectively.

Placemaking with Information

New applications of digital technology have frequently been framed as either tools or publishing media. This is understandable, since other new technologies have often taken the form of tools, and most previous information technologies have been in service to publishing information. However, thinking about these technologies as tools or publications limits our understanding of what they do for us (and to us). They're much more than that.

Consider one of the greatest artifacts to have emerged from the internet: Wikipedia. On one level, Wikipedia is a product: a publication. As suggested by its name, it's modeled on traditional paper-based encyclopedias such as the *Encyclopedia Britannica*. However, Wikipedia differs from those old books in two main ways. First, it's much larger than any previous encyclopedia. As of July 2017, the English version of Wikipedia had 5,435,446 articles and is growing at a rate of over 20,000 per month. To illustrate in physical terms, in 2015 the American artist Michael Mandiberg printed out the English version of Wikipedia. It took up 7,473 volumes, each 700 pages long.[10] The second way in which Wikipedia differs from previous encyclopedias is related to how it got

9. C. Soukup, "Computer-Mediated Communication as a Virtual Third Place: Building Oldenburg's Great Good Places on the World Wide Web," *New Media & Society* 8, no. 3 (2006):421–440.
10. https://en.wikipedia.org/wiki/Wikipedia:Size_of_Wikipedia

so big: Wikipedia is a "living" document that is collectively written and edited by people around the world in real-time. The very instant his printers started putting ink to paper, Mr. Mandiberg's physical copy of Wikipedia was out-of-date.

Astonishing as these figures are, this should be familiar enough ground. After all, we are still talking about a document, even if its size and rate of change happen on a previously unimaginable scale. But on another level, Wikipedia is also the place in which this publication is written. Much as medieval scriptoria provided the ideal environment for monks to hand-copy manuscripts, Wikipedia provides the environment where a small army of mostly anonymous editors and writers can create an organic, networked, decentralized, massive text.

Wikipedia may not have a roof and walls, but it's very much a place. It provides the structures, navigation systems, and rules of engagement that enable over 100,000 people to spend at least an hour every day working there, and for the rest of us to get delightfully lost exploring the many nooks they've created.[11] Wikipedia's founder, Jimmy Wales, is not so much its editor-in-chief, but rather the architect of an environment that made it possible for Wikipedia to emerge from the collective efforts of a large group of globally distributed contributors, most of whom he will never meet. Much like our monumental buildings, Wikipedia-as-place is also laden with meaning as a representative artifact of a new type of culture that works, thrives, and lives in information. No physical place could do these jobs better.

For the most part, the people who design websites and apps have thought of them either primarily as products or services, not as places. While wayfinding has long been part of the digital design discussion, it's been primarily deployed in service for facilitating access to information. With the growing pervasiveness of information systems in our daily lives, placemaking has started to emerge as a primary

11. https://en.wikipedia.org/wiki/Wikipedia_community

concern in the design of information systems. Books such as Malcolm McCullough's *Digital Ground* (2004), Andrea Resmini and Luca Rosati's *Pervasive Information Architecture* (2011), Martin Dade-Robertson's *The Architecture of Information* (2011), and Andrew Hinton's *Understanding Context* (2014) all make compelling cases for consciously crafting contexts with software.

Software-based experiences have become central to our ability to act skillfully. Thinking about them as products, publications, or services is not serving our needs well. If we are to move our shops, schools, singles bars, and third places online, it behooves us to look at how such places have accommodated our needs successfully in the past. Approaching software design as a placemaking activity—with a focus on intended outcomes and behavior rather than on forms or interactions—results in systems that can serve our needs better in the long term. In order to do so, we first need to unpack how environments affect our behavior. That will be the focus of the next chapter.

We shape our buildings;
thereafter they shape us.

—Winston Churchill

2

Context

Take a moment to look around you now and then come back to this paragraph. While your attention is currently focused on these words, your body is located somewhere in space; maybe it's a room in your house or a bench in a park. Your body's relationship with these surroundings will have an important impact on your experience of reading this book. The configuration of the space you're in is either conducive to the task of reading, or it isn't. Is it too loud? Too cold? Is there enough light? Are there other things clamoring for your attention?

You can read more easily in a library than in a bustling nightclub. These two environments create contexts that facilitate very different goals. Your body reacts to cues in these environments in predictable ways: The library encourages you to be quiet and contemplative, a behavior that aids your comprehension of texts, while the nightclub encourages you to socialize.

If you were raised in a culture that has libraries, being in an environment that has the cues you associate with a library will influence your behavior in specific ways. For example, you know what you can expect to do there and what is expected of you as a participant in that type of environment. As designers, we can design these cues: we know some elements and forms lend themselves to reading while others lend themselves to partying.

It's not just physical environments such as libraries and nightclubs that create contexts; information environments create them, too. Just as a library's components make it possible for you to read, the components of a bank's website or mobile application make it possible for you to do your banking. As with the library, the online bank's cues can be designed to create a context conducive to "good banking"—and to put you in the banking mindset—whatever that means for the bank's customers.

Thus, if we want to design information environments that truly serve our needs, we must start by understanding how context works and—more specifically—how we can use language to create particular contexts.

Information architect Andrew Hinton offered a very useful working definition of context in his book *Understanding Context*:

> Context is an agent's understanding of the relationships between the elements of the agent's environment.[1]

In the library example, you are the agent, and the library is the environment. The elements in this environment include the bookshelves, reading tables, chairs, walls, lights, and other accoutrements that make a library a library. These elements are laid out in relation to each other in particular ways—chairs alongside tables that have lights over them, for example—in order to facilitate your use of the place as a context for effective reading.

1. Andrew Hinton, *Understanding Context* (Sebastopol: O'Reilly Media, 2014).

Your understanding of the context of a library is something that you've acquired through previous experiences in such an environment. Babies don't know they're supposed to be quiet in such a place—but you do, perhaps as a result of having been reprimanded by a librarian in the past (or seeing someone else be reprimanded).

I refer to you as an *agent* in the environment because your presence there changes the context. For one thing, you can physically change the form of the environment by moving stuff around. (The librarian may be most displeased!) For another, your mere presence there changes the context. Consider how your experience of being in the library might be affected if you were to suddenly run into Tom Hanks there. (I had this exact experience perusing the aisles in a bookshop in Los Angeles—an encounter that immediately changed my understanding of the context I was in.)

Where Are You and What Can You Do There?

Physical environments—buildings, towns, cities, parks, etc.—are designed artifacts, but we experience these things differently than other designed artifacts, such as iPhones and coffee table books. We experience buildings as urban environments that we inhabit; we move around and inside them, and their forms determine what we can do at any given time.

As we move through an environment, our senses register sights, sounds, smells, and so on. We slowly develop an understanding of the relationship between the different spaces that make up that place. We get a sense of what we can and can't do there. At first, we must rely on our senses and think about what we're doing. Eventually, it becomes second nature.

If you've ever visited a new city, you may have had the experience of being disoriented at first. As you move around, you register particular places in the environment: this is the hotel where I'm staying, one block north is the bakery with the beautiful croissants, two blocks further is the tram station, and so on. Given enough time in the environment, you eventually build a mental representation of the place. You no longer need a physical map to know where you're going, since you've created a sort of internal map of the place.[2] You know where you are relative to other parts of the environment, because you've internalized the parts of the environment and the relationships between them. As a result, you also become more adept at making predictions about what you're likely to find next.

You don't come to this experience as a blank slate. Your expectations of how the place is supposed to be organized are set by your previous experiences and cultural expectations. For example, in a pedestrian-centric European city such as Lyon, you will expect a degree of density and a mix of uses (i.e., commercial and residential) that are quite different than what you'd expect in a car-centric American city such as Houston.

How do you experience a city as pedestrian-friendly? The environment in such a place offers cues that tell you what you can and can't do there. These cues are called *affordances*, a concept introduced by psychologist J. J. Gibson in the 1960s.[3] Gibson and his collaborator and wife Eleanor were interested in how organisms sense their environments. He coined the word *affordance* to describe how elements of an environment communicate the possibilities for action they afford to organisms that are capable of undertaking such actions. For example, to a being with opposable thumbs, a tree branch affords grasping.

Your relationship with your immediate environment—and how you behave in it—is determined by the affordances it provides. Pause for a moment to examine your current demeanor in the environment. You're

2. Psychologists call this a *cognitive map*, a term introduced by Edward Tolman in "Cognitive Maps in Rats and Men," *Psychological Review* 55, no. 4 (1948): 189–208.
3. James J. Gibson, *The Senses Considered as Perceptual Systems* (New York: Houghton Mifflin Company, 1966).

probably holding this book (whether paper-based or in an electronic device) while sitting in a chair or couch in a room of some sort. As an artifact, the book has certain characteristics that make it evident as to how it may be manipulated. You can pick it up, turn it around, and put it inside another object (such as a bag). The same goes for the chair: its form communicates to you that it's ready to receive your butt. It does this by having a particular shape, a particular height, particular materials, and a particular surface treatment that make it adequate for a being such as yourself to sit on.

Imagine how different things would be if the chair's seat were located 11 feet off the ground, or if it were covered in electrified spikes. In such cases, it would not afford "seating" to you. This is an important point: affordances are not inherent characteristics of objects. They only pertain to the relationship between an object and an agent in the environment. A chair that affords seating to you provides completely different affordances to an *E. coli* bacterium. To the bacterium—a microscopic organism with a completely different mechanical configuration and sensory apparatus—a chair does not afford seating.

It's important to note that affordances don't tell you what the book is about; they merely tell you it's an object that you can pick up and manipulate in particular ways. The book provides much information beyond this. For example, its cover may feature its name and the name of the author prominently, a designed feature that comes in very handy when trying to select a particular book from a bookshelf. The information conveyed by the book's cover is an example of a signifier, "some sort of indicator, some signal in the physical or social world that can be interpreted meaningfully" in Don Norman's definition.[4]

When you're walking on a sidewalk in a pedestrian-friendly city like Lyon, you perceive affordances and signifiers all around you. For

4. D. Norman, "Signifiers, Not Affordances," published in ACM *Interactions*, volume 15, issue 6 http://www.jnd.org/dn.mss/signifiers_not_affordances.html

example, the sidewalk itself affords you travel in particular directions at a safe distance from the large, fast-moving objects on the nearby road. The sidewalk doesn't necessarily convey any meaning beyond "you can walk here." There will come a point where the sidewalk ends, and you must cross a road in order to continue on your walk. The road has signals that tell the drivers of vehicles when they should stop, and tell you and your fellow pedestrians when you can walk safely across the road. These crossing signs are signifiers: they convey meaning to both drivers and pedestrians that influence their behavior in the environment.

How You Know What You Can Do There

The meaning of these signs must be learned. We aren't born knowing that red means stop and green means go; these are social conventions we must internalize if they are to communicate their intended meaning to us. And knowing what the colors of the lights mean is not enough: we must also understand the social hierarchies and functional objectives of the environment these colors are enabling. For example, green and red lights have a different meaning on a Christmas tree than they do on a traffic light.

A useful framework for understanding how this works was postulated by media theorist Neil Postman. Postman argued that effective communication required a shared understanding of the social relations between the agents that participated in an interaction, their goals in the interaction, and the particular vocabulary they used when interacting. He called this set of conditions "the *semantic environment* the agents were operating in."[5]

---

5. Neil Postman, *Crazy Talk, Stupid Talk: How We Defeat Ourselves by the Way We Talk and What to Do About It* (New York: Dell Publishing, 1976).

For example, think of the differences between science and religion. You participate in either field to pursue different goals: furthering your understanding of the natural world in the case of science and enhancing you spiritual development in the case of religion. You pursue these goals by using particular social constructs (the priesthood/layperson hierarchy in the case of religion and the peer review process in the case of science) and specialized vocabulary (the language of prayer and scripture in the case of religion and the taxonomies of particular disciplines in the case of science). Science and religion are two areas of human interaction that create and employ different semantic environments.

For you to understand what is going on in any situation—for the communication to make sense—you must abide by the norms of the semantic environment that you're operating in. Attempting to use a religious approach and terminology while performing scientific research would result in bad science. The semantic environment of science allows you to use language to pursue the goals of science effectively by constraining you to a particular context. Your agreement to abide by these constraints is what makes it possible for meaningful communication to happen in this context, and for science to happen at all.

Often, these constraints are implicit and must be learned, as in science and religion. However, sometimes they are explicitly stated. Consider speed limits: there is nothing physically constraining you from driving as fast as you can down any particular road. But society has collectively agreed that some constraints are necessary in order to share the roads (goal) safely, so there are signs (a particular type of vocabulary) in the physical environment to tell you what the speed limit is for the particular stretch of road you're in. If you exceed the stated limit, you run the risk of being ticketed by a law enforcement officer (social hierarchy). Thus, speed limits create a semantic environment that you use to interact safely in the roadways.

Nothing about this sign physically constrains the driver from going too fast; its role is purely semantic.

IMAGE BY RAYSONHO @ OPEN GRID SCHEDULER / GRID ENGINE - OWN WORK, CC0, HTTPS://COMMONS.WIKIMEDIA.ORG/W/INDEX.PHP?CURID=39983160

Software, too, creates a semantic environment. When somebody opens a software application, she does so because she has a particular use (goal) in mind. The application also employs a particular vocabulary that has special meanings in that context. For example, imagine someone (let's call her Lillian) opens Microsoft Word because she wants to write a document (goal). When Lillian opens the application, she sees a primary navigation bar that presents the following choices:

Microsoft Word's primary navigation bar.

The words in this user interface—Home, Insert, Design, Layout, etc.— have particular meanings when used within Word. Lillian knows that "Home" here does not refer to her home in the real world. While its use in this context may not be entirely obvious to her at first (after all, "home" is a fairly generic term), she knows that in this case it's being used in a way that is particular to Microsoft Word—even if this is the first time she has even opened the app. The same is true for all the other words in the navigation bar.

Word's primary navigation system also includes icons. As with the words above them, these, too, have specialized meanings when used in this context. Many of them may be familiar to Lillian, but that's only because she's encountered them before, either in previous versions of Word or in similar applications. These icons have specific meanings when used in this context. One of the challenges a new user of Word faces is learning the meanings of these things in this particular con-text. ("I wonder what this button does? Oh, I see—that means right align.") The set of words, phrases, icons, and other semantic elements in Microsoft Word's user interface creates a semantic environment that should make it possible for Lillian to achieve her goal of writing a document. How much instruction she will require to learn the partic-ularities of the environment depends on many factors, including her level of experience with similar applications.

Microsoft Word is a general-purpose application; its potential user base is anyone who needs to write something. That's a very broad remit! Because of this, Word's designers need to be careful with the language they select so that it's common enough to be broadly understood, yet particular enough so that users know what they can do in the various parts of the application. Other applications have narrower audiences. For example, I once worked on the design of a software system that served as a marketplace for buyers and sellers of energy in regional electricity grids. This was a job with a very particular vocabulary that was only meaningful to the people who worked in this industry. Terms and acronyms that might have been completely baffling to you and me were obvious to these people when used in this context. Since the software that would support their goals would not be used by a general audience, the semantic environment manifested in its user interface leveraged the industry's specialized language to reduce the new user's learning curve.

As a mental exercise, try examining a website or app's navigation system in the absence of company names or logos. How much do the navigation systems tell you about what the place is? How does this change what you understand them to mean?

Here's an example:

I've covered the logos on this website so that you can't tell which company it is. (Although if you live in the U.S., you may be able to guess from the colors.) Look at the words on the navigation bars. You don't need to know anything else about this environment to guess that you're in a bank. One of the labels even says it outright: "Banking." This changes the meaning of the other words there. For example, the word "Learning" could point to many things. However, knowing that you're in a bank helps you constrain the possible meanings of "learning" to something like "educational material for becoming more financially savvy."

To summarize, the words you use in the navigation systems and headings of websites not only help you find what you're looking *for*, but they also help you understand what you're looking *at*. This particular group of words set in this particular order creates a context that gives the meaning to the whole picture. They tell you where you are and what you can do there.

You Are Here

Let's stop again for another moment. Focus your attention on your surroundings and then come back to this paragraph. Hopefully by now it's clear how the environment you're in creates a context that is conducive to your activity of reading. (Or perhaps it doesn't—in either case, you're part of a context that is affecting your behavior.)

While you don't experience them as physical places, websites and apps are also environments. The user interface of a word processor creates a context that affects how you think about what you can do within it, much as a church or a library does. This context is a semantic environment that influences your thinking and behavior. When you're working in a word processor, processing words is what you do. Those words end up in a document that you save in a filesystem, which is another semantic environment that has been established by a software user interface. It's context all the way down!

Whether you're designing a physical environment (such as a church) or an information environment (such as a word processor), you must be aware that you are creating a context that will affect how its users behave in it. The success of the design depends on whether or not it supports the goals its users have for the sort of place it creates. But there is another important set of goals the environment must accommodate: those of its creators. Lillian wants to be able to easily curate a document, and Microsoft wants to continue to generate revenues from providing access to its software. Successful environments bring these goals into balance. The next chapter examines what motivates people to behave in specific ways in their environments and what the incentives are that lead them to design environments to influence those behaviors.

It is difficult to get a man to understand something, when his salary depends on his not understanding it.

—Upton Sinclair

CHAPTER

3

Incentives

The next time you walk into a bank branch—even an old one, like One Montgomery—take some time to consider your behavior. You speak a certain way when you're in a bank. You walk to certain areas in the space and not others. If there are other people waiting to be helped, you wait in line. Why do you act like this when you're there? Does the organization of the environment have anything to do with your behavior? The answer to this question is yes: the environments you inhabit have an important effect on your behavior.

Since this is the case, it's important to examine why you use and create different types of environments. You go to bank branches for different reasons than those that compel you to visit football stadiums. Banking and sporting events fulfill different social needs, and these needs require different ways of acting. Our species has developed particular types of places that influence our behavior in ways that are conducive to meeting these social needs. Underlying it all are incentives that motivate you to act in particular ways.

For example, if you're like most bank customers, your goal in using the bank branch is to do your banking there as quickly and efficiently as possible and then leave. If you behave with civility while there, following directions and standing in line patiently awaiting your turn, you are rewarded with prompt service. However, if you cause a ruckus, whooping and hollering (something that wouldn't be unseemly in the stadium) or jumping the line, you may get in trouble. And if you attempt to enter certain parts of the environment (e.g., the vault) without permission, you may be arrested. Both of these latter outcomes are at odds with your goal of being done with your transaction as quickly as possible.

On the flip side, the bank is also subject to incentives. If it doesn't offer you a particular level of service—for example, if you have to stand in line for too long—you may switch to a competitor. Thus, it is in the bank's best interest to encourage certain behaviors and discourage others. In controlling the environment where the interaction happens, the bank has a great deal of influence in how you behave while you're there.

Let's look at this example in more detail. The internal layout of many bank branches imposes a separation between the bank's tellers and its customers. A barrier (in the form of the counter) keeps the two groups physically separated. On the teller side of the counter, the layout of the space is organized to allow multiple tellers to focus on individual customers and their needs. If you peek behind the counter, you will see cubicles, each with the necessary equipment (a computer, printer, money counting machines, etc.) for an individual teller to serve a customer's requests. On the client side of the counter, space is set up to encourage queuing, often by means of a simple tape barrier, and in some cases by using a virtual line such as a log book.

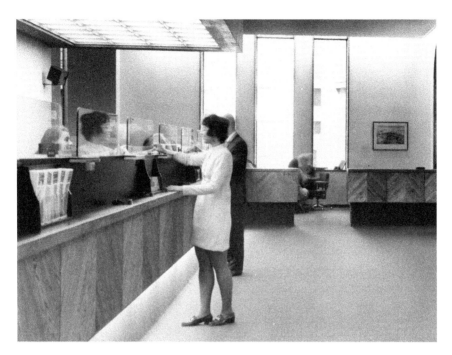

A typical bank branch interior encourages customers to queue so that they can be served individually by tellers who sit in stations behind a barrier.

PHOTO: HTTP://WWW.LOC.GOV/PICTURES/COLLECTION/HH/ITEM/MA0445.PHOTOS.076398P/

The bank's internal layout encourages a particular process. One teller deals with the needs of one customer at a time, although there may be multiple tellers available at any one time. Customers form a single queue, *and* the next customer in line walks up to the next available teller. This process has emerged over time as an efficient way to provide many banking services in a limited physical space.

The word *effective* here means "to satisfy the basic needs of the most amount of customers in the least amount of time by the least amount of tellers." If a customer's need is not basic or routine, such as opening a new account, he is usually directed to another part of the office. There he can start a different process, which the environment accommodates in different ways. For example, the customer may be directed to a small office where he and a bank officer can sit and process the necessary paperwork.

You could imagine a hypothetical bank in which every single customer who walked in—regardless of their need—would get highly personalized services without having to wait in line. This might require employing a larger number of bank tellers. The space of this imaginary bank would have to be laid out differently, since it wouldn't require queuing. It would also require more back-of-the-office space for the larger number of tellers required. In short, the incentives that drive customers to use environments—and the incentives that drive the people who manage them—are a powerful force that determines how environments are structured.

Types of Incentives

There are many ways to incentivize people. Let's examine three of them:

- Remunerative incentives
- Social incentives
- Coercive incentives

Remunerative incentives are those that reward you by offering you something in return for a particular behavior. Remunerative incentives are often monetary, but they don't necessarily have to be. For example, you're rewarded for observing your place in the line at the bank by saving time.

Social incentives are those that affect your social status or self-esteem. For example, trying to cut in line in the bank may earn you scornful looks from other customers; if you know any of them, you may be concerned about being thought of as a boor.

Coercive incentives are those that punish you when you fail to act in a particular way. For example, if you're waiting in line at the bank and start shouting obscenities while waving a gun above your head, you're likely to get arrested (or even shot).

The different types of incentives are powerful tools to influence behavior, sometimes on the scale of a whole society. Think of how taxes are used to implement particular government policies. In the United States, the tax code allows taxpayers to deduct mortgage interest payments from their tax bill. This is a remunerative incentive (the taxpayer gets to keep more of his or her money) that encourages people to buy a home. On the other hand, if people fail to send their payments by the dates established by the government, they are charged a fine. This is a coercive incentive that encourages them to pay on time.

These incentive structures haven't emerged accidentally; they've been designed to achieve those effects. Alas, incentive structures can also have unintended consequences, so they must be handled with care. As Steve Jobs said, "Incentive structures work. So you have to be very careful of what you incent people to do, because various incentive structures create all sorts of consequences that you can't anticipate."[1]

Factors That Drive Incentives

While incentives are an important factor in determining how you behave, certain conditions need to be present for incentives to be effective. These include the degree of agency people have within the system, the power relationships between people in the environment, the stability of their identity over time, and the degree of transparency the environment affords them. Let's examine them in more detail.

Agency

For incentives to influence your behavior, you must have the freedom and ability to act in some ways and not others. If you can't effectively choose one course of action over another, you can't be incentivized. For example, while it makes sense for a bank manager to incentivize

1. https://www.fastcompany.com/1826869/lost-steve-jobs-tapes

tellers to deal with customer transactions within a set amount of time, it's unreasonable for tellers to do so while hovering six inches over their chairs. Levitation is not a course of action that is open to bank tellers; for a manager to attempt to incentivize them to do so would only result in puzzled (and concerned) looks.

The degree to which you have agency affects your behavior whether you're operating in a physical environment or an information environment. While in the physical environment you are subject to a series of real-world constraints (such as the force of gravity), many of the constraints you face in information environments have been designed into them. For example, Twitter used to place a constraint on the length of posts: they needed to be 140 characters long or less. This incentivized certain types of posts and disincentivized others. The constraint was doubled to 280 characters in 2017, leading to an increase in user engagement and a decrease in post abandonment. It seems that when offered more room to express themselves, Twitter users expressed themselves more.[2]

Power Imbalances

Incentives imply an unequal relationship between parties. The party that sets the incentive holds some degree of power over the party being incentivized. The bank teller could choose to ignore the incentive ("slacking off"), but then her employment would be terminated. The criteria for continued employment is determined by the power structure inherent in the relationship between an employee and his or her management.

Managers, in turn, are also incentivized to behave and perform in certain ways by the owners of the company. (In small organizations,

2. https://www.theverge.com/2018/2/8/16990308/twitter-280-character-tweet-length

managers and owners may be the same party. In large organizations, owners' interests are represented by the company's directors.) Owners, in turn, are also incentivized to behave in particular ways. For example, they may experience a rise in the company's share price, a remunerative incentive set by an entity called the *market*. In some countries, such as the U.S., public company managers have a legal obligation to act in ways that meet particular performance targets. In other words, there are remunerative and coercive incentives for them to act in specific ways.

The people who control the shape of the environment have greater power than those who merely use it. Through their architectural decisions, the bank's managers influence the behavior of their customers and employees in the environment. The same is true in an information environment such as Twitter. For example, the decision to expand the length of posts to 280 characters was one that only Twitter's management and their agents could implement; end users merely abided by the decision.

Identity

Most of us want to be perceived by others as being good citizens, parents, or friends, etc. People who live in small communities where everybody knows everybody else have a vested interest in maintaining others' esteem of them. "Old-fashioned" ideals such as keeping true to your word are based on the notion that "you" are a somewhat constant entity over time. In other words, having a stable identity within a community is a key component for social incentives to be effective.

When you interact with others in the "real world," your senses allow you to witness the words and actions of other people directly. "I heard your insult," you say, or "I saw you push him." When identity can be fluid, people are less motivated by social incentives. In Twitter's information environment, for example, participants can be anonymous and create or eliminate accounts on a whim. In such environments, people are shielded from the long-term social implications of their behavior, and do not face coercive (punitive) or moral/social incentives for their negative actions. Anonymity affords them impunity for behavior that wouldn't be tolerated in physical environments.[3]

The ability to permit anonymous accounts is a design decision. Twitter has been dealing with the consequences of this decision by providing means for people to report abusive behavior and block perpetrators. Unfortunately, these mechanisms place the onus of policing behavior on the victims. Creating an environment that mandates stable identities over time would make identity more valuable, curtailing potentially obnoxious behavior. Note this doesn't necessarily mandate that identity in the information environment should be tied to identity in the physical environment. This is an important point to keep in mind, since users may be living in societies where free speech is not tolerated, or where revealing too much may otherwise place them at risk.

Transparency

In this case, *transparency* means easy access to the information necessary for any one individual to know how he or she is doing with regard to the incentives he or she is being measured by. Let's return to the bank teller for a moment. If the teller's bonus is dependent on how many customers he's able to serve per month, the bank should provide him with one critical piece of information—how many customers he's served thus far for any given month—and keep that information updated.

---

3. A good model for this middle path is one of the earliest online communities: the WELL. Even though the WELL was a free-wheeling place where almost anything was acceptable, nobody was allowed to be anonymous. One of its founders, Stewart Brand, coined a phrase that described the community's approach: "You Own Your Own Words" (YOYOW). In other words, having fixed identities over time meant users had an incentive to take responsibility for their actions within the system.

Imagine the bank teller's bonus is contingent on serving more customers than any of his colleagues in any given month. The bank teller may have an understanding of how many customers he's served thus far, but he likely won't be able to have solid numbers on how his colleagues are doing. On the other hand, his managers have a more comprehensive understanding of how the whole team is doing at any given time. They can choose how much (if any) of this information to disclose to the teller, and when.

In physical environments, such as the bank branch, you can often get a sense for how things are going by merely paying attention. While you may not have solid numbers on how many customers one teller is serving at any moment, you might have a notion of how tellers are doing overall by keeping track of how many customers you see in the space. In information environments, transparency is not inherent; rather, it must be designed.

For example, when using Twitter, you have access to a variety of measures that allow you to get a sense of the popularity of your posts. You can see how many people have "liked" or "retweeted" your posts, or how many people have followed you. These are measures that provide positive incentives: as a Twitter user, you are driven to improve all three of these measures. However, Twitter does not provide you with a measure of how many users have "unfollowed" you in a given period of time, a coercive incentive that is not currently being exploited by the system. Given the amount of information that can be obtained from user behavior within a system such as Twitter, asymmetrical access to information about the activity of the various parties involved is potentially much greater in an information environment than in a physical one.

Aligned Incentives

It's important to realize that while the design team (and the broader organization) has goals and incentives that drive them, so do the people who use the environment. Sometimes these two sets of goals are in alignment—that is, they work for each other's benefit. Unfortunately, sometimes they aren't. Let's start by looking at an example where the goals of the people managing the environment are aligned with those of the people using it.

I have two credit cards with Chase bank. During my time as a Chase customer, I've not yet set foot in a physical branch of the bank. I interact with the company exclusively through its iOS apps and website.

When you first visit the bank's website, you will see a screen that looks like this:

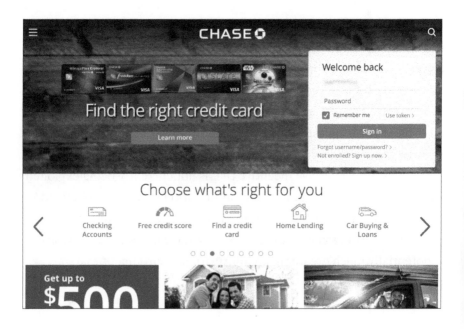

In this screen, you're not shown all that the bank has to offer. Decisions have been made on your behalf to nudge you in particular ways. The most prominent elements on the screen are a series of invitations to learn about their products and a login box. As with many other bank websites, this one is divided into two main areas: a public area that anyone can enter (this is where all the products are), and a private area where you can see your account information. You need a special key—a username/password pair—to enter the private part of the environment.

This page serves two groups of people: those who are already Chase customers and those who aren't. The login box is what most of us Chase customers probably come here looking for, which is why the designers of this screen have made it so prominent. You don't have to click on a "login" button to see the form—the fields are right there, waiting for you to enter your username and password. It's obvious in the same way that the queue is obvious in the physical bank branch.

Once you log in to Chase's website, its navigation changes:

This part of the environment only serves one group of people: Chase customers. You can now move between three areas: accounts, investments, and payments and transfers. Why these three and not others? The people who designed this website clearly considered these to be the categories that address the most important or frequent tasks that you, as a bank customer, need to do there.

If they're like most other banks I've worked with, Chase's website team tracks their customers' behavior in this environment very carefully. The team is incentivized to improve their customers' experience of using the site. On the flip side, an environment that makes it easy for customers to accomplish their goals incentivizes them to prefer Chase over its competitors. It's a virtuous cycle generated by the alignment between the team's and the customers' goals.

Misaligned Incentives

Sometimes the incentives that drive the people using the environment aren't aligned with those that drive the people who manage it. This can result in frustrating experiences for the people who use the environment and ethical conundrums for the people charged with shaping it. Imagine a bank that decides to incentivize its personnel to convince customers to spend more time at the branch, perhaps so they can be sold additional services. A customer who visits such a bank expecting to transact quickly and inexpensively will probably not be very happy. On the flip side, bank personnel will be put in situations that probably make them uncomfortable, trying to convince customers to stay against their best interests.

As a customer, you participate in the bank's environment with the tacit understanding that there is some degree of alignment (or lack thereof) between your goals and those of the people who created and manage the environment. You know that in this particular space, you and the bank's managers and personnel are not peers. But you also understand how that power imbalance works toward the goals you want to accomplish by participating in that shared space.

Places such as banks serve important needs in our societies, and the spatial nudges they use to influence us help us accomplish our goals while also meeting the goals of the people who run these places. Because we've operated in physical environments for such a long time, we've internalized the power dynamics inherent in these places. Information environments, on the other hand, are still relatively new. For the most part, we haven't even thought of them as contexts at all.

We lack the sophistication as users (or as designers) to understand when the interests of the parties involved are misaligned. As a result, we're in the process of moving key social functions to environments driven by incentive structures that are sometimes at odds with the social needs they aim to fulfill. In the words of one former Facebook executive, "We have created tools that are ripping apart the social fabric of how society works."[4]

A big part of the problem is that we've been thinking about these things as tools and not as places. The most serious—and most pervasive—of these misalignments has to do with what people in technology euphemistically call "engagement." That will be the focus of the next chapter.

---

4. https://www.theverge.com/2017/12/11/16761016/former-facebook-exec-ripping-apart-society

We know that we are continually subjected to a huge range of sensory inputs and internal experiences of sensations and thoughts. In fact, almost anything existing in our universe, that can come into human and other animals' purview, can be experienced as information.

—Marcia Bates

CHAPTER

4

Engagement

You walk into the kitchen with the intent of making a
sandwich, when suddenly you hear glass shatter. You
immediately turn toward the source of the sound. Your pulse
quickens as scenarios play in your mind. Has someone broken into your house? Where are your kids? You walk into the
living room to discover your son with a surprised look on his
face and a ball lying on the floor next to the shattered window.
Fortunately, he's alright. You comfort him and discuss what
has happened, and then take your phone out and Google gla-
ziers. You find a company that seems reputable and call them
to set an appointment for the next morning. You go back to the
kitchen and wonder, "Now, where was I?"

Thus far in this book, we've been discussing tangible ways in which places influence our behavior. But there are also more subtle ways in which environments affect us. One that is of particular importance is how they impact our ability to focus our attention.

Sometimes our attention is taken away by an exceptional occurrence, such as the sound of a breaking window. This is useful; the ability to respond quickly to changing conditions can help us escape danger. However, most of the time, we want to be in control of our attention. An environment that nudges us to spend more of our time there—or keeps interrupting us—would make it difficult for us to get things done.

The places we inhabit can either allow us to remain in control of our attention or snatch it from us for purposes of their own. Unfortunately, many of today's most popular information environments are based on business models that incentivize the latter. The term used in the technology business is "engagement": the amount of time people spend looking at or interacting with components in the environment. Given how important our attention is, it's worth looking at how designing for engagement affects it.

What Attention Is and Why It Matters

You can think of attention as your ability to focus your mind on one piece of information among many so that you can achieve a particular goal. The sound of breaking glass offers your senses new information that interrupts your train of thought. It causes you to suspend your immediate aim—the sandwich—in favor of another, more urgent one: making sure everything's okay with your home.

As you read this paragraph, you're sensing information about your environment: the temperature of the space, various background noises, the level of lighting, and so on (including the words that make up the paragraph). Your mind is also prompting you with information unrelated to the words you're reading: memories of what you had for breakfast, a plan for this evening's date, a reminder to call your mom, and so on. Your ability to finish reading this paragraph requires that you somehow ignore these distractions so that you can focus on the running stream of words your eyes are making available to you.

This ability to focus our minds is an essential survival mechanism. Our remote ancestors wouldn't have lasted long in the savanna had they not been able to look out for predators. Their survival required that they pay close attention to their surroundings for new pieces of information: a rustling in the grass, a particular musky smell, and so on. On the flip side, it would have been impossible for our forebears to hunt if their minds kept getting caught on whatever their eyes and ears happened to land on at any given moment; they needed to remain alert. Survival required that they be able to marshal their cognitive resources toward particular goals (e.g., steak!) to the exclusion of others (e.g., crane flapping overhead!).

Note how attention is closely related to the environment. Even though our mental chatter is among the pieces of information we need to sift amongst, much of what we focus our attention on are stimuli outside of us, conveyed to our minds by our senses. That musky predator smell is not emanating from our bodies (well, not from most of us) but from something out there in the world. We learn to recognize these stimuli and tell them from each other: for example, this noise hearkens food, this one signals possible death, this one means it's safe to take a nap, and so on.[1]

---

1. For a more in-depth summary of how attention works and how it is affected by technology, see Adam Gazzaley and Larry D. Rosen, *The Distracted Mind: Ancient Brains in a High-Tech World* (Cambridge: MIT Press, 2016).

Foraging for Information

As you can see, the ability to focus our attention toward particular goals is an essential part of our biological makeup. Given how central these mechanisms are to our experience of the world, it's worth delving a bit more into how they evolved. Obtaining food is an essential goal for an animal. Alas, foraging does not come free: scouting the environment in a focused way requires that the animal expend lots of energy. The animal must take in more energy than it uses if it is to survive. Over time, evolution has selected patterns for foraging that allow organisms to acquire the highest energy sources at the lowest energy costs. Optimal foraging theory is a model that predicts how animals will behave when they search for food in this manner. It takes into consideration factors related to the structure of the animal itself, such as its size and metabolism, its ability to move around and carry food with it (and hence, the time required for it to search for food), and environmental factors such as the distances between possible sources of food.

In the late 1990s, researchers working at Xerox's PARC (Palo Alto Research Center) noticed similarities between the way people search for information and the way that animals forage for food. The theory they developed, called *information foraging*, has been influential in the history of user interface design. The basis of these theories is the idea that our ability to focus our attention developed as a survival mechanism, which made it possible for our ancestors to find food in a complex environment while avoiding getting killed.[2] While it's unlikely you'll meet a hungry predator on your way to the water cooler, you still inhabit complex environments which offer all sorts of distractions. Whether it be making it across a busy intersection or finding an email from your landlord, this ancient ability to exclude some pieces of information in your environment in favor of others is essential for you to

---

2. P. Pirolli and S. Card, "Information Foraging," *Psychological Review* 106, no. 4 (1999): 643–675, https://pdfs.semanticscholar.org/59f1/e446fab201399d84ae48156ea7bd05c4439b.pdf

accomplish your goals. Understanding how our attentional mechanisms work helps us design environments that allow us to achieve our goals more by cutting down on distractions.

Being There

What does it mean for you to "be" somewhere? Let's examine where you are right now. Are you sitting in a room while reading this book? Perhaps you're out-of-doors, in a park or natural setting. Whatever the case, your body is located somewhere in space. You can feel the effects of gravity holding you down.[3] If you choose to focus your attention, you can feel the points where your legs touch the chair or your feet touch the floor, and the weight of the book or tablet as you hold it. You can sense the temperature and humidity of the air around you. Your body can only be in one place at a time, and, wherever that is, there you are.

But is that really where "you" are? After all, your body is not all there is to you; you also have a mind that is experiencing all of this. This mind is not only the apparatus that allows you to focus on some of these external stimuli by ignoring others, but it also provides distractions of its own. For example, at this point, you may be thinking, "What is he going on about? Why did he add that stupid footnote about the space station? Why should I care about this stuff?" (Who said that? Where does that voice come from? Do you consciously decide to pay heed to it? Or does the chatter start of its own volition? Whose volition?)

"Being" somewhere—being you—is the experience of a particular stream of stimuli (both external and internal) that you choose to focus on, consciously or not. You can think of it as a sort of spotlight that allows you to illuminate different things at any given moment.[4] Right now, your attention is on these words. Two paragraphs ago, it was the

---

3. Unless you're reading this on board the International Space Station, in which case I salute you.
4. Adam Gazzaley and Larry D. Rosen, *The Distracted Mind: Ancient Brains in a High-Tech World* (Cambridge: MIT Press, 2016).

pull of gravity on your body. This shifting of attention between various pieces of information becomes the content of your life. You make decisions based on the stream of things you focus on. Over time, this stream leads you to develop a particular way of understanding and being in the world. As José Ortega y Gassett said, "Tell me what you pay attention to, and I will tell you who you are."

At the risk of depressing you, this is where I must remind you that at some point this stream will stop. Humans have an average lifespan of 79 years. (How far into this number are you? I'm keenly aware of being well past the midway point myself.) Whatever happens to your consciousness after death is a matter for religious debate, and we will not get into that discussion here. What is indisputable is that in this plane of being, the time available for you to experience the world is limited.

When we focus intensely on something or someone, we say we are paying attention to it. This metaphor suggests that we think of our attention as a sort of resource or currency that we spend at will. However, this idea of attention-as-currency is not accurate, since we can't save our attention for later expenditure. The current moment is all we have, and once it's gone, it's gone forever. If we're to think of attention as a resource, it should be as one that is nonrenewable. Your attention is one of your most precious possessions—one you should zealously guard from squandering.

Buying and Selling Attention

Part of the reason why attention is so precious is that what you attend to influences how you understand the world and thus, how you behave. Because attention is so limited and so essential to our survival, we're taught to handle it with care. You may recall Aesop's fable about the boy who cried wolf: the boy kept calling his neighbors' attention, tricking them into thinking a wolf was coming to attack his flock of sheep.

When the wolf eventually showed up, nobody believed the boy, and he and his flock were eaten.

If I can claim your attention, I can change your behavior. If there is, in fact, a predator hiding around the corner from the water cooler, knowing so in advance could mean the difference between enjoying a drink of water and exiting the gene pool prematurely. If I possess this piece of information about the state of the environment and you don't, I have an advantage over you. I can choose whether to share this information with you or not. This understanding of the current state of the world has a material effect on your decision-making abilities: if your goal is to get a drink without dying, knowing a predator is on the prowl can make the difference.

Gathering information about the state of the world is not without cost: it may be that I'm risking my life by being on the lookout for predators, or that I've invested in installing devices that allow me to get a read on their presence. I also incur costs in sharing that information with you. For significant parts of human history, people who wanted to leverage other's knowledge of the state of the environment had to pay for the privilege by exchanging their own information, or later buying a book or newspaper. In groups larger than small hunter-gatherer societies, access to information did not come cheap.

In the first half of the 19th century, we reached an important milestone in making access to information more accessible. A New York publisher called Benjamin Day had a brilliant idea: he would sell a newspaper to the masses below what it cost him to produce it. Instead of charging fees for people to access the information in the paper, his revenue would come from selling advertising in the paper's pages. He based this innovative business model on the premise that selling access to information to the people who wanted to use it would be less lucrative than selling access to those people's attention to third parties who wanted to influence their behavior.

Day's paper—*The Sun*—became the first successful advertising-supported mass medium in the world, ushering in a new age in which the common man could access the news of the day. In order to attract readers, much of *The Sun's* content leaned toward the baser aspects of human experience: suicides, violent crimes, petty courtroom drama. The paper also fabricated fantastic stories, including the famous Great Moon Hoax of 1835, which falsely attributed to renowned astronomer John Herschel the discovery of—among many other outrageous things—a civilization of humanoid bats on the moon.

Illustration of a moon scene described in the August 28, 1835 edition of *The Sun*.
IMAGE: HTTPS://WWW.LOC.GOV/RESOURCE/PGA.02667/

Sensational content produced sensational results: *The Sun* soon became very popular, and advertising became established as an effective means for generating revenue by bringing awareness of products and services to potential customers. In a newly industrialized world, which produced a surplus of goods, the ability to harvest attention to create demand became essential. This surplus included products such as patent medicines, which owed much of their success to newly created techniques such as "expert" endorsements and the promotion of exotic ingredients such as snake oil.

During the first half of the 20th century, the emerging techniques of mass attention harvesting were used to create demand for all sorts of other products, from automobiles to cigarettes to conscripts for two world wars. As new media appeared on the scene (radio, television), they were co-opted from their initially altruistic goals toward the harvesting of their audience's attention as the primary business model. The internet proved to be no different. When it became popular during the final decade of the century, entrepreneurs investing in the new medium were primed to consider advertising as a revenue source.[5] (I remember the pre-advertising web. Seeing my first ad online felt like a shocking intrusion into what had thus far been a noncommercial experience.)

But attention harvesting in digital information environments is not the same as it was in prior media. For one thing, advertising in broadcast media such as newspapers, radio, and television is a blunt instrument. Everyone in the audience must sit through ads that are of little or no relevance to them. Online advertisers can build rich profiles of individual users and follow them as they move through various websites, tracking the effectiveness of individual ads over time. As a result, the "right" ad can appear to the right person at the right time.

Interactive information environments also offer advertisers the ability to close the sales cycle without the user having to leave the medium. The user of a website can find out about a new product online, educate herself about its benefits, and purchase it in one session without having to get up from her chair. And, of course, another significant difference is the one explored by the premise of this book: interactive information environments create

5. For an excellent overview of the history of advertising, see *The Attention Merchants* by Tim Wu (New York: Vintage Books, 2016).

contexts in which people engage with each other and with their social, governmental, and business institutions. These contexts can change and adapt as they learn more about their users' behaviors and what stimulates them. This combination of factors makes information environments incredibly effective mechanisms for persuasion.

A contemporary "boy-who-cried-wolf" moment.
The designer of this advertisement has included a hair
on the image to get us to swipe on it.
SOURCE: HTTPS://TWITTER.COM/BLAKEIR/STATUS/939605849340895237

Toward What End?

The Sun deemed it appropriate to mislead its readers into believing the moon was populated by exotic creatures, because its business model rewarded engagement over elucidation. The more that people paid attention to the paper, the more money it made—regardless of why they chose to do so. In this case, the incentives that drove the various parties were misaligned. On the one hand, the paper's readers wanted to be accurately informed about the state of the world, and they expended their energy by focusing their attention on the pages of *The Sun* with that goal in mind. On the other hand, the publishers of the information received more money the more of this attention they attracted.

We may laugh at the naïveté of the 19th-century public, but this misalignment of economic incentives is still very much with us. Although both publishers and audiences have become more sophisticated, most contemporary media still rely on business models that monetize the audience's attention. And today's information environments don't just present information about the state of the world—as discussed previously, these places are also the context where much discussion, learning, and decision-making take place. Our attention is increasingly focused on these information environments not so that we can act more skillfully in physical environments, but because information environments themselves are where we work, play, learn, and build our societies.

When we enter an information environment with the expectation that we're there to participate in a community, and the business model that supports the place seeks to maximize our engagement, our goals and those of the people who manage the environment are working at odds with each other. The writer Zeynep Tufekci has used a powerful analogy to describe the design of information environments that monetize their users' attention:

"[the site] has to keep poking you to get you to stay on the site. I liken this to holding meetings in a place designed to keep meetings going forever.... If a person is in a meeting, what is it you want? You want it to end.... Whereas these places are designed to make sure meetings never end. And that's where we're holding our public sphere, in places that are designed to keep us engaged by making us go down the rabbit hole."[6]

The more time you spend in these places, the more of your attention they command. Because selling your attention is their primary source of revenue, they will do everything they can to keep you engaged. Tufekci notes that extreme positions are better drivers of engagement than more nuanced perspectives, and environments that use algorithms that learn from user data patterns to select the most engaging content will tend to push users toward fringe positions.

An environment that captures its users' attention by radicalizing them in this way will impact their ability to participate in reasonable discourse, especially if it becomes their primary source of information about the world. Ironically, the effects of pushing toward engagement may also be detrimental to advertising itself, since no brand that values customer trust wants to appear to support extreme positions, conspiracy theories, or sensationalism. In 2017, Procter & Gamble—the world's largest advertiser—announced that it was curtailing its investment in digital advertising in part because of what's euphemistically referred to as "brand safety": the unwillingness of corporations to continue having their brands associated with the sort of fringe viewpoints that can emerge from engagement-based models.[7]

Some of the most important information environments in the world currently rely on selling their users' attention. For example, Facebook

6. https://samharris.org/podcasts/persuasion-and-control/
7. http://adage.com/article/cmo-strategy/p-g-slashe/309936/

has over two billion monthly users,[8] and its primary source of revenue is selling their attention to advertisers. This situation would not be problematic if Facebook's goals were trivial, but the company does not think of itself as a provider of mere entertainment. Its stated mission is to "Give people the power to build community and bring the world closer together."[9] In other words, Facebook aspires to become the third place for the world.

As we saw in Chapter 1, third places serve a particular role in our society: they are where we catch up with our neighbors, where we get a sense for the pulse of politics and culture, and where we belong. But an engaged citizenry is not necessarily an informed one, and, in fact, may be quite the opposite. For a third place to function effectively as such, we must be able to enter and leave it at will. We must also be clear on what the price of admission is.

Consider another social institution, which has been explicitly designed to function as a third place: Starbucks. The Starbucks website explains its origin story thus:

> In 1983, [Starbucks founder Howard Schultz] traveled to Italy and became captivated with Italian coffee bars and the romance of the coffee experience. He had a vision to bring the Italian coffeehouse tradition back to the United States. A place for conversation and a sense of community. A third place between work and home.... From the beginning, Starbucks set out to be a different kind of company. One that not only celebrated coffee and the rich tradition, but that also brought a feeling of connection.[10]

8. https://newsroom.fb.com/news/2017/06/two-billion-people-coming-together-on-facebook/
9. https://www.forbes.com/sites/kathleenchaykowski/2017/06/22/mark-zuckerberg-gives-facebook-a-new-mission/#59117c091343
10. https://www.starbucks.com/about-us/company-information

Students at a Starbucks in Singapore.
IMAGE: HTTPS://COMMONS.WIKIMEDIA.ORG/WIKI/FILE:STUDYING_IN_STARBUCKS.JPG

Visit any regular Starbucks store in the late morning, and you'll get a sense for how successful this vision has been: You'll find people there working on their notebook computers, reading, or sitting in small groups chatting. These people don't go to Starbucks merely because of the coffee, but because it provides them with a space to focus on their work, their leisure, or each other. The cost for using this place is obvious to its customers: all they must do is buy a beverage or food item. In exchange for their money, customers get a cup of coffee and the use of Starbucks's facilities for a while. The ambiance is casual; for a small price, people can stay there as long as they like. (Of course, the longer they linger, the more likely it is they'll make further purchases.)

Now imagine a competitor that took on Starbucks by offering the coffee and food for free or below cost, much like *The Sun* did in its time. To make money, this new chain of coffee houses—let's call it Brewhaha— would charge other companies to show its patrons ads while they're in the store. While similar to Starbucks on the surface, Brewhaha would be a very different place. Every once in a while, you'd be interrupted and asked to focus on some new piece of information. Brewhaha would be incentivized to work more and more of these interruptions into your attention stream and to keep you there as long as possible so that you could see more and more of them. It would also be incentivized to do more unsavory things, such as co-opting your friends to serve as representatives of the companies paying it to advertise.

Even if these ads were highly targeted to your interests and were unobtrusive, Brewhaha would not be a very good place to communicate with others. It would be tough, for example, to have a sustained, focused conversation with a business prospect in such an environment. The neutrality of the place as a container for such conversations would be in doubt. You'd suspect the environment was studying you and nudging you in particular ways to keep you engaged and focused on things other than what you came there for—and you'd be right.

Your understanding of your role in the transaction would also be less clear since you're no longer handing over cash for the right to be there. The experience would appear to be "free," even though, in fact, it cost you a much dearer price than money: your ability to remain focused on your goals. (This is not to say Starbucks is devoid of advertising; most Starbucks stores feature ads for their own products and for some third parties. However, advertising is not Starbucks's primary source of revenue. Once you buy refreshments from them, you're left to your own business in the space.)

Former Google Design Ethicist Tristan Harris has documented ways in which technology is being used to manipulate people to drive engagement. These include the use of psychological techniques such as intermittent variable rewards, which are also used to design addictive experiences, such as slot machines. When you feel compelled to compulsively check your social media status, you're doing so by design.[11]

---

11. https://journal.thriveglobal.com/how-technology-hijacks-peoples-minds-from-a-magician-and-google-s-design-ethicist-56d62ef5edf3

Because these information environments are funded by advertising, the people who design and operate them are incentivized to keep you there and to keep you engaged; to check your status—and the place's advertisers—just one more time. This wouldn't be as much of a problem if we used these environments with the understanding that we're there as adults to be entertained, as we do with casinos. But if these environments are to become the places where we hold our civic and social discussions, they must move beyond the selling of attention as their primary line of business.

It's worth noting here that this is not a black-and-white issue; not all advertising-based information environments are necessarily "bad" for us. It comes down to what we want to use the place for, and making sure our goals and the goals of the people operating the environment are in alignment. For example, I have fewer issues with a search engine, such as Google, using an advertising-based business model. This is because I use the search engine to find stuff. Since the advertiser's goal is get me to buy their wares (and this requires that I find them), our goals are more closely aligned than if I were using the place to chat with my friends.

An environment designed to help us engage transparently with each other is very different than one designed to keep us engaged with it. Such "greedy environments" are not conducive to the sort of civil discussion that is essential for healthy societies. Unfortunately, these greedy environments have become incredibly powerful and pervasive. They exert a tremendous influence in forming people's political and social opinions.

As alarming as this may sound, it's only the beginning of the story. New technologies are coming on the scene, which will make today's information environments seem innocuous and transparent in comparison. For most of our history, we've inhabited environments that have been carefully designed to affect our ability to focus our attention in particular ways. For example, the layout of a church asks that you focus your attention on the altar, and not on the other congregants. Like them, the first information environments we've been inhabiting—software, websites, and apps—have also been consciously designed by people. We can examine their form and structure and understand how they act upon us and our ability to focus.

Now, however, we're starting to see information environments whose structures and components are not determined by people, but by data-driven algorithms. These algorithms learn and adapt to optimize their effectiveness, and in so doing reify the incentive structures they were created to serve. The results are not entirely predictable, and the reasoning behind them is opaque. The next chapter explores this and other technological advances that will soon impact our ability to control our attention and therefore our ability to reach our goals.

Technology is the campfire around which we tell our stories.

—Laurie Anderson

CHAPTER

5

Technology

A large crowd looked intently toward the center of the tall glass room, where an open-sided shaft rode up toward the ceiling. A platform slowly made its way up the shaft, pulled up by a taught rope. Standing on top of the platform, a man surveyed his audience. When the platform had reached a high-enough point, the man gave the order and an assistant slashed the rope with an ax. The audience gasped. The platform fell a couple of inches and stopped. The year was 1854, and the man was Elisha Otis, inventor of the elevator safety brake.

Otis's dramatic demonstration at the New York Crystal Palace ushered in a new age for architecture. Elevators had been around for a long time, but they were dangerous contraptions: a malfunction could send passengers plummeting to serious injury or death. As a result, they were used mostly for indus-trial purposes. Otis's invention allowed people to ride safely in long vertical stretches. One result: the skyscraper, made possible by two technological advances of the late 19th century: Otis's safety brake and steel beam construction.

Elisha Otis demonstrating the first elevator safety brake in 1854.
IMAGE: HTTPS://EN.WIKIPEDIA.ORG/WIKI/ELISHA_OTIS#/MEDIA/FILE:ELISHA_OTIS_1854.JPG

The form of our environments is not influenced solely by the goals and incentives that drive us to make them; technology plays a very important role as well. Placemaking and technology have always gone hand-in-hand. New technologies make new kinds of environments possible. Conversely, the social need for new types of environments spurs technological advances. And while this is true of physical environments such as sky-scrapers, it's even more true of information environments like websites and apps, which we experience exclusively through technology.

Information technologies such as transistors, integrated circuits, touchscreens, and cell networks have made it possible for us to create contexts that change how we understand and behave in the world. New technologies bring new capabilities with them, much like Otis's safety brake did. There is much excitement and experimentation when a new technology appears on the scene. Entrepreneurs and tinkerers explore the possibilities with projects that are mostly technology-driven. Eventually, people come to understand the value of the new technology and put it to use for people—not for its own sake, but for what it can do to make their lives better.

Many of the information technologies that underpin our current infor-mation environments are still in their early stages, and there are many more new ones on the horizon. The pace of development in the infor-mation technology space is astonishing, and as a result we still see products and services that seem to be primarily in service of the tech-nologies that make them possible, instead of the people who will use them. Understanding this, and knowing when it's time to move on from the experimentation phase to put technology to use toward meeting human needs instead, will allow us to create information environments that better meet our needs in the long term.

Innovation in information technology happens very fast, so I write these words knowing that this chapter of the book will age less grace-fully than the rest. But if we aim to create information environments that stand the test of time, we need to understand the impact that technologies have on the form and capabilities of these environments and their ability to shape our attention. Looking at the technologies that promise the most change in how we relate to information is a good place to start. In this chapter, we will look at four such technologies.

Reframing Reality

Let's start with the most obvious: virtual reality (VR) and augmented reality (AR). I say "obvious" because when I told people I was writing a book about information environments, many assumed it was about VR or AR. While the principles laid out in this book certainly apply to VR and AR, they also apply to other currently more common ways in which we access information environments today.

In the short history of digital computing, we've accessed information environments mostly through words and pictures on computer screens and interacted with these words and images using keyboards, mice, and various less conventional devices such as graphics tablets and light pens. Since the introduction of the iPhone in 2007, touchscreens have also become popular.

All of these interaction mechanisms—from the earliest "dumb" computer terminals to the latest tablets—have one thing in common: they show us information inside glass rectangles that we access intentionally and which we can distinguish from the "real" world. I'm typing these words on a MacBook Pro. The laptop's screen is a few centimeters from my face; a black bezel around the display delineates the boundary between the information environment where I'm writing this book and the physical environment where I can access my coffee mug. I consciously switch my attention between the two. I type a sentence in one, then take a sip in the other, with no ambiguity about which is which. VR and AR change this relationship in fundamental ways.

VR is a set of technologies that replace the signals you receive with your senses from the physical environment with signals entirely generated by the computer. Instead of seeing information inside a glass rectangle "out there" in the world, VR replaces your field of view and sound with one generated by the computer. Sensors keep track of your body's position and geometry in physical space so that the simulation can adjust the perspective in real time. If you rotate your head slightly, the image of the simulated environment moves accordingly.

Augmented reality seems similar to VR on the surface, but is, in fact, a different proposition. Instead of aiming to entirely replace our environmental cues with simulated ones, AR overlays computer-generated cues over those we get from the "real" world. For example, using AR, you could conjure a three-dimensional model of a house on top of your dining room table. Moving your body around the table would allow you to see all sides of the model as though it were there "in real life." Using similar tracking technology as that used in VR, the simulated images are kept synchronized with your field of view, making it seem as though they are really there.

While the applications for AR are less obvious than those for VR, I believe it is the more powerful of the two technologies. The ability to replace your reality with a completely different one can be enticing for entertainment or research, but being able to overlay digital information over physical environments opens up possibilities in a variety of fields. For example, the assembly of a complex product such as an aircraft could be made easier if instructions were overlaid on top of the physical parts in the plant.

AR and VR blur the lines between digital information environments and physical environments. Instead of being something you access within the confines of a computer display, digital information becomes an integral part of your experience in the world. (In the case of VR, it becomes your experience of the world altogether.) Having a constant layer of information meaningfully overlaid over the "real world" is still the stuff of science fiction, but we're not far from achieving something close to it.

Having access to constant real-time ambient information about things in the world would radically transform our interactions. Imagine being able to see information about another person overlaid when you meet them, instantly and unobtrusively; you'd never again be at a loss for an acquaintance's name, for example. This capability opens up great possibilities for a better understanding of the world, other people, and ourselves. But it also suggests dystopian scenarios in which governments and corporations mediate our experience of reality to an even greater degree than they already do. And, of course, wearers of such augmentation technology would have an unsurmountable edge over nonwearers. In either case, we can expect the boundary between physical environments and information environments to become blurred as these technologies improve. The potential for behavior manipulation through context- and goal-setting implied by the blending of physical and information environments is unlike anything we've experienced before.[1]

Machines That Decide for Us

One of our species' longest-standing dreams is to create machines able to make sense of the world and act independently from us. In Greek mythology, Hephaestus—the god of blacksmiths and craftsmen—created metal automatons to help out in his workshop and do his bidding, and the sculptor Pygmalion fell in love (and had children) with an ivory statue he made. More recently, movies such as *2001: A Space Odyssey* and *Her* offer visions of a world in which we co-exist with nonhuman entities of our own devising.

The advent of digital computers has allowed us to move closer to this goal. As an area of research, artificial intelligence (AI) has been around since the mid-1950s, and the field has made progress in fits and starts since then. However, recent advances in our ability to acquire, store, and process data have led to a flourishing of new AI applications. Computers are not (yet) smart enough for us to fall in love with them,

---

1. For a deeper discussion of the blending of physical and information spaces, see "Designing Cross-Channel Ecosystems as Blended Spaces," by Andrea Resmini and David Benyon, https://blogs.aalto.fi/multixd/files/2016/09/MultiXD16_paper_9.pdf

but they can diagnose diseases and drive in busy urban environments without crashing.[2]

Current AI systems simulate reasoning by drawing inferences and making decisions from patterns they perceive in data. They acquire this data by being fed symbolic databases (as with Watson, the Jeopardy-beating system developed by IBM), by detecting cues from the environment (as with LIDAR-wielding autonomous cars), or both. The net result is software that can fly airplanes, translate texts from one language to another, beat humans at complex strategy games such as *Go* and chess, recognize and label the content in photographs, and more.

These systems don't need to mimic human interactions (like the HAL 9000 computer in *2001: A Space Odyssey*) in order to be useful. In fact, you may not even be aware of their work behind-the-scenes. Consider how Amazon.com's homepage shows you products you may be interested in based on your past shopping and browsing behavior. This is a simple AI that examines the types of products you've recently purchased or browsed for and shows you more of the same. Rather than working as an agent you interact with, the Amazon algorithm works like an architect, dynamically changing the form of the environment especially for you.

Recommended for You in Kindle Books

Amazon's algorithms are smart enough to suggest books for me to read based on other books I've bought there. However, they're not smart enough to know I've already read one of these books through Audible, an Amazon subsidiary.

2. People working in the field of artificial intelligence distinguish between "strong" AI—the possibility of creating systems that exhibit general intelligence—and "weak" AI, which is focused on narrow tasks. We currently have the latter.

Having access to an information environment that is custom-tailored to your needs and desires can make it easier for you to find the things you're looking for. It can also drive revenue by allowing customers to find new products and services they wouldn't have discovered otherwise.

The suggestions you see when you log into Amazon are mostly right for you and would be of less interest to somebody else. Amazon is not an environment where you meet with friends or participate in a community. However, there are other environments in which community-building is a primary goal. In these cases, giving everyone access to different information may be counterproductive.

Consider the case of Facebook's news feed (the list of posts you see when you log into the environment). The news feed is not a simple chronologically ordered list of the latest stuff your friends have published. Instead, the posts you see and the order they're shown in are curated by algorithms that learn from your behavior and interests over time. These algorithms rearchitect key aspects of the Facebook environment every time you log in, with one objective in mind: to keep you engaged.

This has two primary effects. The first is that no two Facebook news feeds are identical. If you start a conversation within such a feed, your perspectives will be informed by the different contexts you're operating in. You may be seeing things which upset you, and which are invisible to me. This may sound far-fetched, but consider that Facebook has conducted at least one psychological experiment in which they manipulated the emotions of their users by altering the amount of positive or negative posts they were shown in the news feed.[3]

The second effect is that only seeing things that keep you engaged can lead you to develop a distorted perspective of reality. For example, the algorithm may decide only to show you posts by people who agree with

3. V . Goel, "Facebook Tinkers with Users' Emotions in News Feed Experiment, Stirring Outcry," *New York Times,* June 29, 2014, https://www.nytimes.com/2014/06/30/technology/facebook-tinkers-with-users-emotions-in-news-feed-experiment-stirring-outcry.html

you politically. This leads to the effect known as a "filter bubble," in which you come to believe your views are universally held because you only see posts that agree with your views.

Both effects are pernicious to healthy societies. Resilience calls for open dialogue between people with diverse points of view, and unstable contexts that steer our conversations for the sake of engagement are unable to host such dialogues in an unbiased way.

Machines That Speak Our Language

Not all AI algorithms operate quietly behind the scenes. Advances in voice recognition and language processing have led to a surge in "smart assistants," such as Apple's Siri, Microsoft's Cortana, and Amazon's Alexa. These smart assistants can recognize our requests in (somewhat) natural language and respond to inquiries about current weather conditions and song requests, or even control the lights in our living room. Other conversational AIs, such as x.ai's Amy help us coordinate a meeting or book a hotel room through established channels like email instead of through a voice-driven interface.

Early computers required that their users communicate with them by typing arcane commands into a text-based command line. The next dominant paradigm, graphical user interfaces (GUIs) such as the ones used by Apple's Macintosh and Microsoft Windows, allowed users to interact with computers by pointing at menus and icons. Current touchscreen-based smartphones and tablets offer yet another paradigm: touchscreens that enable users to (literally) reach out and manipulate information with their fingers. As advanced and "learnable" as these interfaces are, they still require more thinking than regular conversation does. Interacting with other people by talking is something our species has done for much longer than we've had computers. Speech is our universal interface to our social world.

There are many challenges to overcome when creating computers that can understand natural speech and respond in kind. Parsing sentences is relatively easy, but recognizing intent isn't. There's much subtlety in our spoken language. Words can have multiple meanings. Context is important and so is intonation. ("Is he asking a question or making a statement?") Advances in AI have enabled the current crop of "intelligent" digital assistants to act upon simple English requests, such as "Siri, remind me to email Hans about the proposal."

Conversational user interfaces change how we experience information environments in essential ways. For one, when we rely solely on spoken commands, we can't see available options like we can with visual user interfaces. Eventually, "smart" agents such as Siri and Alexa may become clever enough to engage with us in unstructured conversation, but for now, interacting with them requires thinking through how we phrase and pace our queries. For another, having an open microphone always listening for a trigger phrase ("Hey, Siri!") changes our perceptions of the boundaries between physical and information environments. When your living room becomes part of a computer's user interface, you must become more conscious about what you say and do there.

IMAGE: XKCD (HTTPS://XKCD.COM/1807/)

Connected Things Everywhere

Over time, computers have been getting more powerful, cheaper, compact, and energy-efficient. One result of these trends is that they've started to "break out of the box" and make their way into all sorts of everyday objects and appliances, ranging from light bulbs to jet engines.

Why would you want a light bulb with a computer in it? Because you can do things with it that you can't do using regular light bulbs. For example, my home's front door is illuminated with a Philips Hue light bulb, one such "smart" device. I've configured this light to turn on automatically at sundown and turn off at sunrise. It knows when to do so because it's connected to a weather service over the internet that provides it with the exact times for sunrise and sunset in my location. I can also turn this light on or off (or in various gradations in-between) remotely from my smartphone, or configure it to come on when it rains.

My light bulb is a one-way device. I can only program and control its behavior remotely, and it doesn't know anything about its surroundings. However, many of these devices also have sensors in them that allow them to act on signals they receive from the environment. For example, the Nest thermostat senses the temperature in your house and adjusts itself accordingly. Using machine-learning algorithms, devices such as this thermostat can act on a wide variety of data sources—gleaned from their sensors, third-party services over the internet, and from historical and real-time data generated by other such devices. The result is devices that automate everyday tasks in much smarter—and often more efficient—ways than were possible with the "dumb" objects of the past.

A subcategory of these "Internet of Things" (IoT) devices is what is often called "wearable" computers. These are devices designed to be worn unobtrusively by people to augment their abilities. For example, the Apple Watch helps monitor my body's energy expenditure and heart rate. It also notifies me when messages arrive and unlocks my computer when I get near it, obviating the need for a password.

IoT devices transform our relationship with information in various ways. To start, they provide ambient (or peripheral) access to information. Previously, we had to pull out a smartphone from our pocket or open our laptop computer if we wanted to enter an information environment. It was a conscious decision. However, with IoT devices around us, dynamic digital information can become part of our physical environments. We no longer have to explicitly choose to switch our attention to computers and smartphones in order to access information; now it's out there, as part of everyday objects in the world, where it's more difficult for us to miss.

Also, IoT devices can act upon the physical environment. My Hue light bulb illuminates a part of the physical world, not some simulation. A smart lock can change the state of your front door in a very tangible way, either permitting or forbidding entrance to your home. A smart thermostat can adjust the temperature of your house based on how many people are currently occupying the space, the outside temperature, and a host of other variables.

Finally, IoT devices have sensors that gather data about the state of physical environments and their inhabitants. I have one such device in my home, a smart camera, which constantly monitors air quality in my home. If we're not home (which it knows because of the fact my wife's and my iPhones are not nearby), it also captures video and audio that allows us to see what's going on in the house. And if its video feed shows movement, the camera triggers a notification so I can check in. The camera is sending this information back to its manufacturer's servers, where it's potentially building a profile of my family's daily patterns.

Is there something nefarious about this? I don't have reason to believe so. But the amount of data that can be gleaned from a device that is continuously monitoring physical space is tremendous, and we've started to surround ourselves with many such devices. As a result, companies will be able to form very accurate data-driven pictures of our private lives.

Disintermediating Trust

Many of our day-to-day activities require that we trust each other. As we saw in Chapter 3, having a stable identity is key to our social interactions; it allows us to determine whether or not a person we're dealing with can be trusted. In physical environments, trust is relatively easy to achieve when dealing with people we're in constant contact with, such as our colleagues and officemates. We keep track of their actions over time. But what about people we don't know? How do we develop enough trust in them to allow us to transact with each other? The best answer we've come up with thus far is to delegate trust to third parties such as banks and governments.

Consider a simple example: you walk into a drugstore to buy a stick of gum. You pick up the gum, walk to the counter, and hand the cashier a five-dollar bill; the cashier hands you your change, and you walk out with your product and a lengthy receipt. You and the cashier never met before, so you have no reason to trust each other. You can complete the transaction because you're both using a form of currency—U.S. dollars—that was issued by a third party (the Federal Reserve Bank) you both trust.

Now let's imagine you're paying with a credit card instead of cash. What happens then? You walk to the counter with your stick of gum, and the cashier asks you to insert your credit card into a POS (point of sale) terminal, a little computer that reads information stored in the card. This information, which identifies this particular card, is sent (along with details about the transaction) from the POS to an organization called a payment processor, who then forwards it to the credit card company (e.g., Visa, American Express, MasterCard, etc.). The credit card company then forwards it to the bank that issued the card to you, which confirms whether it is legitimate (e.g., hasn't been stolen) and has enough funds available. If the card meets these conditions, the bank issues an authorization number, which then makes its way back through the chain: from the bank to the credit card company to the payment processor to the merchant, who then hands you your stick of gum and lengthy receipt. Complicated, no?

What if instead of triggering this elaborate choreography of trust, you and the cashier could trust each other directly, without intermediaries? That's the promise of the blockchain, a new technology that makes it possible for us to check each other's financial bona fides directly, without requiring us to go through banks, credit card companies, or governments.

The blockchain functions like an open—yet anonymized and encrypted—ledger that makes it easy for anyone to consult whether we're good for the promises we make when we transact with each other. While this capability by itself has the potential to change how much of the world works, it doesn't stop there. There's nothing inherent in the technology to limit it to financial dealings; any transaction between two or more parties can be registered as part of a blockchain. As a result, it can replace intermediaries in all sorts of situations that currently call for establishing relationships of trust where none exist.

As you go about your day, consider how many interactions—both on- and offline—depend on trusting the identity of other people, and how many of these interactions are mediated by third parties for the sake of establishing trust between you and the person you're transacting with. The blockchain promises to do away with those intermediaries. If nothing else, having access to someone's history can help us understand whether we are dealing with a trustworthy person or not. The blockchain could enable "instant trust" between people who have never met before without requiring that they reveal their real-world identities. This could create incentives for them to behave in ways that don't impact their online personas negatively. If it were to become widespread, this capability would alter the way we interact in information environments (and in physical environments, too).

Back to the Future

I was eight years old when I first sat in front of a computer. It was the late 1970s, and the machine before me—a TRS-80 Model I—was an ungainly collection of oddly-shaped plastic boxes on top of a desk. I interacted with this computer using a clunky keyboard that wouldn't seem out of place in a 1960s office and a display that looked like a cheap black-and-white tube television. The Model I didn't have much in the way of graphics; the few games in the system used mostly letters and numbers. With a whopping 4Kb of RAM, it couldn't even store today's smallest cat GIF. Still, it gave me many hours of joy as I used it to explore new worlds in text adventures or command the starship *Enterprise*.

Eight-year-old me couldn't imagine the iPhone or Siri, and the web would've blown my mind. (The web still blows my mind—and I've been working on it for over 20 years!) I have three kids who are between five and nine years old as I write this. They take these technologies for granted, much as I took the TRS-80 Model I for granted. By the time my children reach their mid-forties—my age at the time of writing—the means by which they access information environments will be very different than the ones we have today.

We can't accurately forecast what these will be, but we do know some of the things that underpin today's information environments will continue to underpin future ones. It's safe to assume our bodies will have the same constraints they have today. Communication will be essential. Language and context will still play a central role, as will identity and reputation. Information will still need to be findable, understandable, and actionable. Given how central information environments have become to the functioning of modern societies, it behooves us to design them in ways that encourage and support resilience. The next chapter turns our attention to how we can design information environments for the long term.

Architecture can't force people to connect, it can only plan the crossing points, remove barriers, and make the meeting places useful and attractive.

—Denise Scott Brown

6

Architecture

I n 1936, Herbert "Hib" Johnson—head of the S.C. Johnson company and grandson of its founder—hired the architect Frank Lloyd Wright to design a new headquarters for his company. Johnson wanted to "make our chosen life's work fine, and more enjoyable," and to "eliminate the drabness and dullness we so often find in office buildings." In other words, he wanted a better work environment for himself and his employees.

The building Wright designed for the Johnson company is considered to be one of his best. Its centerpiece—a huge, soaring space called the *Great Workroom*—is one of the first open plan office environments. Writing about the Great Workroom many years later, architecture critic Paul Goldberger called it a triumph of engineering and esthetics, a space "created to give the company's clerical workers a sense of community and nobility."[1]

The Great Workroom in the Johnson Wax headquarters.

PHOTO: HTTP://WWW.LOC.GOV/PICTURES/COLLECTION/HH/ITEM/WI0052.PHOTOS.171425P/

I do much of my work with collaborators I seldom see in person. We, too, share a sense of community, but ours doesn't congregate in physical space; our gathering place is a software application called *Slack*. Alas, Slack doesn't do much to reinforce our sense of community—and there is much less nobility.

In Silicon Valley, as well as many enterprises, the default framing for thinking about customer-facing digital things is that they are either

1. http://www.nytimes.com/1987/11/01/arts/architecture-view-wright-s-vision-of-the-civilized-workplace.html?pagewanted=all

products or services. I often meet peers who describe themselves as product designers or (less frequently) service designers. It's not unusual to hear of teams working toward a minimum viable product. The things they're working on have product features that are defined by a product manager. When they launch, these systems are said to be "in production."

Product is an appropriate framing for some classes of digital things— but not all. Slack is not just a product. Android is not just a product. iTunes is not just a product. Facebook is not just a product. Salesforce is not just a product. Weibo is not just a product. These things are also information environments; through their language and structures, they create contexts that change how people understand the world, think, and act. They also serve as platforms where first-, second-, and third-parties can build and host products of their own. The list of their stakeholders is long and extends well beyond the confines of the organizations that "manage" them.

To think of these things as merely products or services is to grossly underestimate the impact they have on the people who use them. We don't experience products or services in a void; we experience them in contexts, and these systems create those contexts. While there's no doubt that you're interacting with a service when you log into your insurance company's website to see details about your policy, that service is being provided to you within a particular context that is created by the website itself. To design the service in isolation from its context is to do neither justice.

As with physical environments, such as buildings and towns, information environments must be designed to address particular needs. Although the design of software is relatively new, the design of environments is not. People have been creating buildings and towns for centuries. The field of design that has focused on designing environments is architecture. In this chapter, we will examine the design of information environments as a form of architecture. Let's start by unpacking what we mean by design.

Design

When many people think of design, they think of aesthetics: how something "looks and feels." They may sense something has "good design" because it stands out from among similar products because of the way it looks. This is a misunderstanding. Design is not a characteristic of things; it's a process by which things are made. It's more of a verb than a noun.

When we're creating something new, design allows us to simulate possible solutions so that we can experience them without incurring the costs of building the final product. We do this by understanding the context the thing we're designing will address (including the needs and expectations of the people who will use it) and producing models of varying fidelity that allow us to envision and test alternatives. For any one challenge, there are many possible ways forward. Design allows us to explore and refine the ones that best serve the needs of the project. Design is how we make possibilities tangible.

These needs are expressed as requirements, which can be overt or tacit. Often, these requirements are at odds with each other, and it is up to the designer to suggest a particular balance between them. I'll illustrate with a case I saw firsthand. Many years ago, I witnessed the design process of an architectural project for a housing subdivision. The subdivision was going to have many houses laid out in long, narrow lots that were perpendicular to the street. An architect had been hired to design the house model that would be built on these lots. He came up with a design that had the roofline of the house set perpendicular to the direction of the lot, so the roof was lower at the front and back of the house and taller in the middle. The architect liked how this looked.

However, these houses were meant to appeal to low-income buyers. This meant that the houses needed to be expandable over time, and the only place for the houses to grow was into the backyard. Having the roofline perpendicular to the lot made that much more difficult than if it were parallel to the lot.

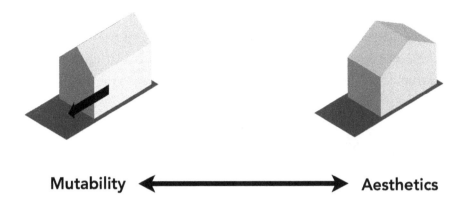

Mutability ⟵⟶ **Aesthetics**

You could say that in this case, two forces were pulling the design in opposite directions. One approach was optimized for resilience and mutability—the ability to change over time—while the other was optimized for the architect's aesthetic preferences.

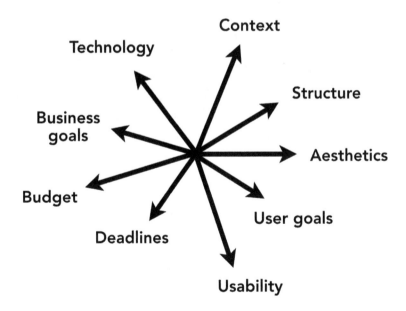

All design projects have forces such as these pulling them in different directions. The design process consists of defining hypotheses of what the balance between these forces ought to be and articulating them to stakeholders—the people who are commissioning it and those who will be using it—and the people who will build it. Designers use models to communicate the intended balance to these audiences. These models can take many forms: sketches, comps, prototypes, etc. As the project proceeds, the models go from more abstract to more concrete, but by definition they always stop short of being of identical fidelity to the final product. Feedback from stakeholders and developers help designers refine the models as the project progresses.

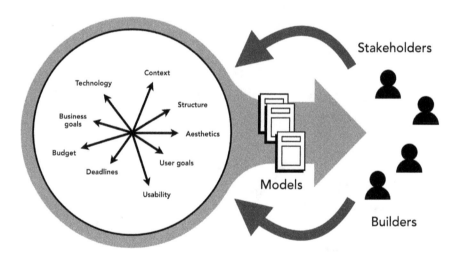

People have used this approach to problem-solving for a long time in a variety of different fields and problem spaces. Particular disciplines have emerged to focus on particular sets of problems. For example, graphic design is the design discipline that focuses on creating visual signs that move us to action. Industrial design focuses instead on creating practical mass produced objects. Architecture is the design discipline focused on the creation of physical environments, such as buildings and towns. Because of this, it is of particular interest to us as we evolve the design of information environments.

The Design of Environments

The first designed environments were very simple; they came about as our forebears experimented with various ways of sheltering themselves, their families, and their goods from the elements. This process of trial-and-error produced a set of best practices, leading to the establishment of a vernacular craft tradition of building that is still alive in many rural areas. But eventually, building design moved beyond the functional basics demanded by shelter to serve other goals, including the creation of a tribal identity and mediation of the group's relationship with the supernatural. Architecture developed a formal language that allowed it to serve these roles above and beyond the mere provision of shelter.

Architecture informs our self-identity and culture more than any other design discipline. This is because the output of architecture revolves around the contexts in which much of our lives play out. Think about the place where you live. Perhaps you live in a house or an apartment. What does this structure say about you? How does it inform your self-image? What do others think about it (and hence, about you)? How intentionally does it reflect who you are? Does it serve merely utilitarian ends, or are there aspects of it that transcend shelter to express something about your individuality or your place in the pecking order of society?

Buildings do more than just protect us from the elements. For example, they can also create contexts that express the wealth and power of their owners. Interior of the Winter Palace in St. Petersburg, Russia.

IMAGE: HTTPS://COMMONS.WIKIMEDIA.ORG/WIKI/FILE%3AHAU._INTERIORS_OF_THE_WINTER_PALACE._THE_BOUDOIR_ OF_EMPRESS_MARIA_ALEXANDROVNA._1861.JPG

Of course, environments don't just affect us at a personal level; they also influence our ability to function socially. For generations, the well or fountain in the town square served as the meeting place where citizens found out what their neighbors were thinking, the guildhall enabled colleagues to work out differences and embark on joint endeavors, and the assembly hall hosted lawmakers as they discussed and developed laws. As we move more of these and other social and civic activities to information environments, it behooves us to vie for their integrity and usefulness. Over the centuries, architects have developed various design principles that can guide us in the pursuit of these goals. Let's examine a few of them.

Function

Architecture is eminently practical. Above all else, buildings provide utility: they keep us warm and dry and give us safe places to sleep, eat, learn, transact, and more. The relationship between form and function has been a central concern of architecture for a long time. In the late 19th century, the architect Louis Sullivan coined the famous phrase "form follows function," which succinctly expresses a focus on efficiency and expression in service to functionality.

Ergonomics

Human bodies have particular characteristics that inform how they interact with the environment. For example, doors must have an appropriate height and width if people are to use them to move from one room to the next; door handles must be located at a specific height if they are to be turned by human hands; and so on. Architects are concerned about designing environments that can be used by people, so they carefully consider their work in relation to the dimensions and capabilities of the human body.

Understandability

Elements in the environment must not only be usable by people, but they must also be *understood* to be usable. The door "reads" as an opening in the wall; it says, "passage here!" Architects have long played with the configuration of elements to communicate particular messages and influence behavior. For example, urban designers will often place prominent vertical elements at junctures to call attention to key nodes in the city's fabric. Successful public environments must be understandable by people from a variety of backgrounds and levels of education.

Urban monuments such as Nelson's Column in London serve as punctuation marks in the environment. They help people orient themselves and move around the environment.

PHOTO BY DAVID CASTOR, PUBLIC DOMAIN, HTTPS://COMMONS.WIKIMEDIA.ORG/W/INDEX.PHP?CURID=6636406

Quality

People experience a beautiful form crafted from fine materials very differently than the same form sloppily thrown together using inadequate materials. Joints between disparate elements and materials can be elegant, inventive, and delightful, or they can be clumsy and thoughtless. Architects strive to infuse their work with a tangible sense of quality from the largest to the most granular scale.

Coherence

The most successful environments have their elements arranged into a particular order that people understand as being more than just a sum of parts. This means that every part of the composition contributes to the coherence of the whole. Most architects strive for gestalt: ensuring elements "carry their weight" conceptually toward a sense of wholeness in the place.

Fit

The form of a designed environment is a response to a set of conditions in the world. The success of the environment depends on how well it addresses the contextual conditions that brought it forth. Architects strive for what Christopher Alexander called "good fit": a tight relationship between the form, or product, we have created, and the context it was created for, with the understanding that the context the project exists in is constantly changing.[2]

Resilience

Environments are constantly adapting to changing conditions outside and inside the environment. Change can come from the external

---

2. Christopher Alexander, *Notes on the Synthesis of Form* (Cambridge: Harvard University Press, 1964).

forces, such as winds buffeting the surface of a building, or from internal forces, such as the need to reconfigure a space to accommodate a new use. Architects are concerned with the adaptability and resilience of the environment; they design environments that can change without compromising the integrity of the whole.

Mental Models

These principles are as applicable to the design of information environments as they are to the design of physical places.[3] Both types of environments have something important in common: people experience them by taking in various parts of the environment one at a time. Eventually, newcomers to a place understand how it's organized and how they can move around in it.

Think back to a time in your life when you were in a new location—perhaps you'd started a job in a new office. Your first days in the new office would be a little disorienting. You'd need help knowing where things were; you probably felt clumsy and got lost often. Eventually, you figured out the "lay of the land" and could move around with ease, without having to think about it. How did you do this?

You did it by developing a mental model of the environment. You built this model by moving around the environment and taking in its features through your perceptual systems. Your memory kept track of where you'd been and created relationships between spaces in your mind. In a physical environment, such as your office, this happens as you walk around. In information environments, you "move" by clicking or tapping on elements such as links and buttons that show you different parts of the environment. In both cases, you are creating an internal model of how elements relate to each other. The more nuanced

---

3. For more heuristics for designing of information environments, see Andrea Resmini and Luca Rosati, *Pervasive Information Architecture: Designing Cross-Channel User Experiences* (Burlington: Morgan Kaufmann, 2011).

this model is, the better you understand the environment, and the easier it will be for you to function there.

Most architects strive to make environments understandable; this entails laying out the place and detailing it so that its users can build mental models of the place easily. Architects do this by carefully arranging relationships between physical forms and elements. Designers of information environments strive for the same objective, but do so by arranging semantic elements such as icons, labels, buttons, and links. These are obviously very different things. Words and pictures on a computer display, or sentences spoken by a "smart agent" in a cylinder on your mantelpiece, have entirely different characteristics than walls and passages in a building. There is an emergent design discipline that does for information environments what architecture does for physical environments. It's called *information architecture*.

The Architecture of Information

Information architecture (IA) is the area of practice and field of study that concerns itself with the design of information environments. IA aims to make information easier for people to find and understand. As with (physical) architecture, this calls for bringing a particular order to the elements that comprise the environment.

Although the structuring of information for ease of retrieval has been around for centuries (think of the invention of page numbers and tables of contents in books), it was with the arrival of digital information technologies such as computers that the discipline of IA came into its own. These two objectives—making information findable and making it understandable—have been part of the discipline since its early days. The emphasis on one or the other has varied over time, but the two have always complemented and supported each other.

As with so many other important contributions to the field of human-computer interaction (HCI), the modern conception of IA traces its roots to groups working at IBM and Xerox's Palo Alto Research Center (PARC) in the late 1960s and early 1970s. Researchers and inventors at these companies were among the first to meet the informational challenges posed by new digital technologies, which made it possible for people to interact with data in scales previously unimagined.

The term *information architecture* itself was popularized by the architect Richard Saul Wurman in the mid-1970s. Wurman's approach was to make things more understandable by presenting information in a particular context that made it relatable on a gut level. At first, Wurman's attention was focused on urban environments. For example, an early book, *Cities: Comparison of Form and Scale*, consisted of photographs of 1:14,000 scale clay models of cities around the world; the first to-scale comparison of cities. Subsequent books (over 80 of them) have clarified a wide range of different subjects, from the work of his mentor Louis Kahn to personal investing to dog care.

Many of Wurman's most well-known projects are either books (*Information Anxiety*, *Information Architects*, the *Access* series, etc.) or maps (e.g., the Tokyo transit map). However, there is more to his work than graphic design. Among the general public, he is better known as the founder of the TED conferences. As with his books and visual design, TED originated with an impulse to clarify by subtraction and reconfiguration; in this case subtracting all the things Wurman hated about conferences and carefully curating speakers from a wide variety of different fields.

The common thread that underlies all of these projects is the idea that the relationships between elements creates a particular context that affects how we understand the whole. In the case of the Tokyo subway map, the elements are the names of individual stations in the system. At TED, they are presentations by speakers from widely different backgrounds that have been carefully curated. As Wurman put it, "You only understand information relative to what you already understand."[4]

4. Richard Saul Wurman, *Information Anxiety 2* (Indianapolis: Que, 2000).

Tokyo transit map by Richard Saul Wurman.

Contemporary information architecture also has roots in Information Sciences, mostly through the work of Peter Morville and Louis Rosenfeld.[5] Their 1998 book *Information Architecture for the World Wide Web,* affectionately known as the *polar bear book* because of the illustration on its cover, brought traditional principles of information organization to the design of websites at a time when many designers were first starting to grapple with the challenges of large-scale digital information environments. It was the right message at the right time, and the polar bear book was very successful.

5. Rosenfeld is the founder of Rosenfeld Media, the publisher of this book.

Architecture was often used metaphorically by practitioners, who understood wayfinding as a goal in both physical and information environments. Christina Wodtke's *Information Architecture: Blueprints for the Web* (2002) made the connection between architecture and web design explicit. The explosion of information that resulted from the popularization of the World Wide Web ushered in a decade in which the IA community focused on finding their way to information. However, when smartphones started to take over as the primary means with which people accessed information environments, information architects shifted their attention to context-making. Andrea Resmini and Luca Rosati's landmark *Pervasive Information Architecture* (2011) was a call for a more systemic approach to design, with the aim of creating coherent experiences that span multiple touchpoints across physical and digital space. Andrew Hinton's *Understanding Context* (2014) tackled the subject of placemaking with language head-on, and is required reading on the subject. Both books provide solid foundations for an approach to architecture that aims to make information not just findable, but understandable. In 2015, the fourth edition of the polar bear book (now titled *Information Architecture: For the Web and Beyond*, and of which I'm a co-author) introduced new material on context-making with information to the classic IA text.[6]

These advances acknowledge that information is only understood in particular contexts, and that the way we organize the semantic structures of our information products and services alters those contexts. It's a perspective that acknowledges information environments as places that should serve the needs of the people who will use them. Unfortunately, it's the opposite of today's deeply cynical and pervasive focus on engagement, which, in practice, seeks to apply our new information superpowers to manipulating and addicting people.

6. For a more comprehensive look at the history of information architecture, see "A Brief History of Information Architecture" by Andrea Resmini, *The Journal of Information Architecture* issue 3, no. 2 (2011): 33–46, http://journalofia.org/volume3/issue2/03-resmini/

Community and Nobility

The largest information technology companies in Silicon Valley—Google, Facebook, Apple—understand the importance of designing environments that enable their people to do their best work. All three have hired world-class architects to design their corporate headquarters: Bjarke Ingels and Thomas Heatherwick (Google), Frank Gehry (Facebook), and Norman Foster (Apple). Like Hib Johnson before them, the leaders of these companies seek to create places that give their workers a sense of community and nobility.[7] Unfortunately, this perspective hasn't yet extended to the people who will be using the companies' products and services, perhaps because designers and managers in these companies understand them primarily as products and services.

Shortly before I moved to the San Francisco Bay Area in 2013, a friend warned me: information architecture is barely a thing here anymore. To my dismay, I've found that to be true. Design here has focused on a product-oriented (and less frequently, service-oriented) approach that seeks to "move fast and break things," a mantra that until recently decorated the walls at Facebook's headquarters. But when your "product" is used for hours on end by billions of people every month, the thing you risk breaking is society itself. It's time we recognize that these digital things we're making are the places where many of our most important social interactions are happening, and start designing them accordingly. These things need architecture.

Over the course of the next three chapters, we will look more closely at how information architects tackle the challenge of designing information environments that serve people's needs. We will start by looking more closely at the role of structure in creating coherent contexts.

---

7. And perhaps also signal to the world that they have the wherewithal to change the environment in powerful ways.

Our capacity to experience, make, and communicate (share) meaning is not just a result of the makeup of our brains and bodies, but depends equally on the ways our environments are structured.

—Mark L. Johnson

7

Structure

It's a crisp evening in the late spring of 2017, and I'm sitting in Berkeley's Hearst Greek Theater along with my friend Alex and our wives. We're enjoying a concert by Jean-Michel Jarre—the "godfather of electronic music"—who is playing his synthesizers and dancing energetically, surrounded by sophisticated electronic instruments and enormous video screens that slide before and behind him. Even though the four of us enjoy each other's company, our attention is squarely focused on the performers. This control of our attention is by design: the layout of the theater encourages us to focus on what is happening onstage.

The structure of our environments has a substantial effect on our ability to perceive information. An environment designed to enable a large audience to pay attention to a performance, such as the Hearst Theater, will be structured very differently than one meant for an individual to find and study a book, such as a library.

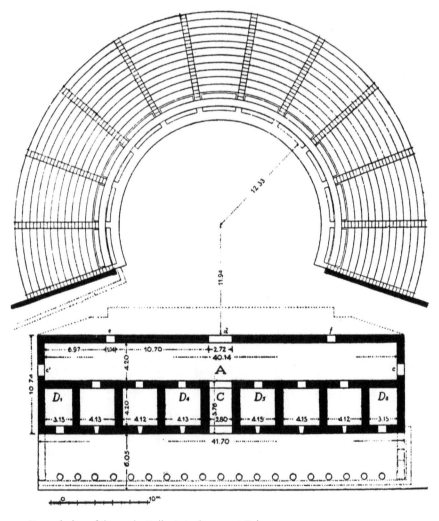

Ground plan of the early Hellenistic theater at Ephesus.

IMAGE VIA WIKIMEDIA COMMONS, HTTPS://COMMONS.WIKIMEDIA.ORG/WIKI/FILE:THE_GREEK_THEATER_AND_ITS_DRAMA_(1918)_(14782041714).JPG

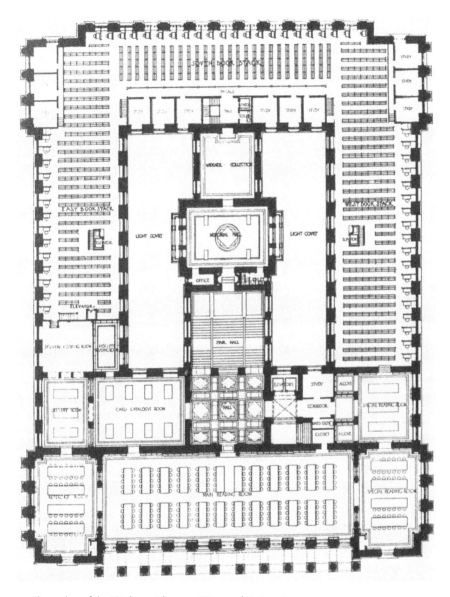

Floor plan of the Widener Library at Harvard University.

IMAGE BY HORACE TRUMBAUER & ASSOCIATES [PUBLIC DOMAIN], VIA WIKIMEDIA COMMONS, HTTPS://COMMONS
.WIKIMEDIA.ORG/WIKI/FILE%3AHARVARDUNIVERSITY_WIDENERLIBRARY_SECONDFLOORPLAN_SNEADIRONWORKS.JPG

There is no one right way to organize environments to accomplish these goals, but rather many different ways optimized toward different ends. We see this in the variety of library and theater designs in the world. However, some arrangements have proven to be most effective over time. We've inherited the structure of the Hearst Theater from the ancient world. This arrangement—with graded seating for an audience set in a concentric circle around a stage—is optimal if you want the audience to pay attention to a performance.

The structural configurations that underlie these places are critical to their ability to serve our needs. They're also essential to the way information environments work. Let's examine how they do this.

What We Mean by Structure (and Why It Matters)

The key to understanding structure is recognizing that we perceive things as being composed of interrelated elements: parts that make a whole. For example, a company may have departments such as marketing, sales, product management, finance, and human resources. These departments have different roles within the organization, and function in a variety of ways, but they all ultimately work together to serve the whole. These elements of the company relate to each other in specific ways that allow them to accomplish their functions by establishing particular relationships between them. For example, while the marketing and sales departments may be peers in one company, marketing may report to sales in another. The company's leadership chooses one arrangement over another, ideally in service to particular business objectives. The set of relationships between the departments (the parts) of the company (the whole) is its structure.

There are various ways to organize the parts. For example, many organizations are arranged in a hierarchy, as is often the case of marketing reporting to sales. Hierarchies are characterized by one-to-many relationships between the parts, and can be visualized as trees, with the highest-ranking element in the structure being the equivalent of the trunk supporting various levels of branches and leaves below it.

Another common structural arrangement is the network. In a network, any part can have a relationship with any other part. While not very common in organizations, there have been some experiments in networked corporate structures that allow teams to self-organize. (The foremost example, online shoe retailer Zappos, is based on a model called *holacracy* that aims to empower its constituent parts to define themselves and their relationships with other parts on an ad-hoc basis.[1])

A third way of organizing elements is a *lattice*: an arrangement in which parts have formal relationships with other parts, which are somehow adjacent to them. Unlike a hierarchy, whose relationships are top-down, relationships in a lattice can be established "sideways" between peer elements. Unlike networked arrangements, relationships in a lattice are not entirely arbitrary; they implement a particular pattern. In companies, this most often manifests itself as "matrixed" arrangements in which groups have both a top-down managerial structure and horizontal responsibilities and relationships with peer teams.

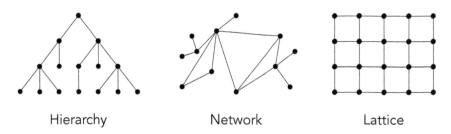

Hierarchy   Network   Lattice

Three types of structural arrangements.

1. https://www.zapposinsights.com/about/holacracy

These various arrangements of parts that make a whole are evident both in the natural world and in the things we make. We don't see them as arbitrary or haphazard; on the contrary, we perceive (and use) *patterns* in these arrangements. Think of the "double helix" structure of DNA, the leaves of a fern, the sequence of sounds and silences in a sonata, and the arrangement of load-bearing components of a building: all are recognizable as having patterns of order.

Ferns exhibit clear organizational patterns.
IMAGE BY SANJAY ACH VIA WIKIMEDIA COMMONS, HTTPS://COMMONS.WIKIMEDIA.ORG/W/INDEX.PHP?CURID=2169955

TEMPLE of DIANA, EPHESUS.

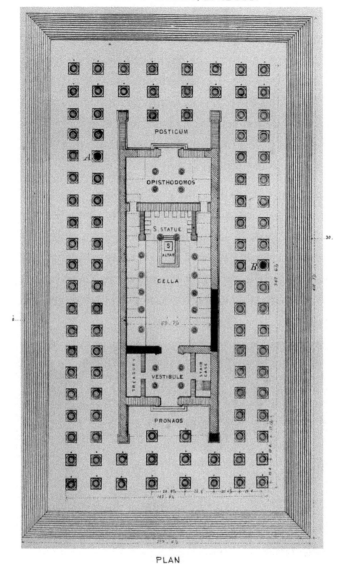

PLAN

SCALE of FEET.

N. B. The Columns marked A & B. and the Walling and Anta colored
dark were found in position.
The dotted Columns are Sculptured. (Columnæ Cælatæ")
The foundation piers of the Church within the Walls of the Temple are

Floor plan of the Temple of Artemis at Ephesus by John Turtle Wood.

HTTP://ONLINE.MQ.EDU.AU/PUB/ACANSCAE/CHAPTERS/CHAPTER09.HTM, PUBLIC DOMAIN,
HTTPS://COMMONS.WIKIMEDIA.ORG/W/INDEX.PHP?CURID=3324874

Pattern-matching is an innate ability in humans which allows us to inter-act more effectively with our environments. We look around and listen and are then able to anticipate the location and behavior of food sources and predators in the environment. Doing this well gives individuals an advantage, so we've evolved to use pattern-matching to understand what we're perceiving (and where it is in relation to ourselves) when mov-ing around in an environment. As a result, it's easier for us to navigate environments that have a clear structure than those that are chaotic and random. It's also easier for us to design and build them.

Many buildings have structures that establish a sense of order and regu-larity in the environment. A beautiful example of this is the Kimbell Art Museum in Fort Worth, Texas, by the architect Louis Kahn. The Kimbell is organized as a series of parallel vaults especially designed to gently shed natural light on the museum's art collection. These vaults establish a regular pattern that makes it easy for visitors to understand the place.

The Kimbell Art Museum is structured as a series of regular vaults.
IMAGE: HTTPS://COMMONS.WIKIMEDIA.ORG/WIKI/FILE:KIMBELL_ART_MUSEUM_HIGHSMITH.JPG

When talking about a building such as the Kimbell, we can discuss structure at least on two levels. The most obvious is the physical (load bearing) structure that makes it possible for the building to resist the force of gravity. This includes elements such as columns and beams arranged in regular patterns; the bulk of the Kimbell's weight is carried by columns on the corners of each of the vaults and by the vaults themselves. The physical structure is what most of us think of when we think about structure in buildings.

However, on another level—one that is more relevant to our topic—we can also talk about the building's *conceptual* structure. In the case of the Kimbell, this conceptual structure is the fact that the building is composed of modules (the vaults) that have particular characteristics; that these vaults are organized in a particular way that creates symmetry and order; that this arrangement establishes particular relationships between outdoor and indoor spaces; and that the various programmatic needs of the buildings (e.g., it must include galleries, a cafeteria, a gift shop, and auditorium, etc.) are served within these vaults, and so on.[2]

In buildings, as well as in information environments, the conceptual structure serves to create both distinctions and coherence between the environment's constituent elements. If it's to succeed, the environment must resolve the natural tension that exists between these two directions. In the Kimbell, distinctions are the various types of spaces called for by the building's program, which it must have to function as such— just as a company must have various departments fulfilling different roles. An auditorium has very different spatial requirements than a cafeteria or a gallery, yet the Kimbell suggests that somehow they must all be served by these vault modules.

---

2. For more details on the Kimbell's load-bearing and conceptual structures, see
   https://www.kimbellart.org/architecture/kahn-building

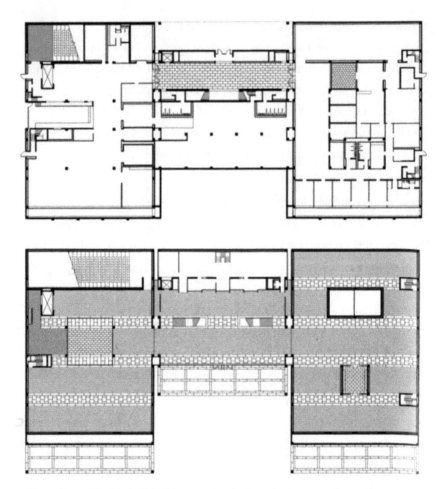

Floor plans of the Kimbell Art Museum clearly show the structure of its vaults.
IMAGE: HTTP://WWW.ARCHDAILY.COM/123761/AD-CLASSICS-KIMBELL-ART-MUSEUM-LOUIS-KAHN/52737DFCE8E44EE8E1000800-AD-CLASSICS-KIMBELL-ART-MUSEUM-LOUIS-KAHN-FLOOR-PLANS

Coherence comes through the use of a common element (the vault) to implement these programmatic functions, and through a consistent level of *quality* in the implementation of the design as a *real building in the physical world*. It's one thing to say "the building is a series of vaults that let in natural light" and another to create an actual building that does this. For the building to function as such, the architect must address many implementation details. For example, how do visitors move from one vault to another? How do transitions happen from public spaces (such as the lobby), through more private spaces (the galleries), to those that are inaccessible to the public (back office)? How do the joints work where one material (concrete) meets another (marble)? How does such a space maintain a comfortable level of temperature, humidity, and lighting, while still allowing the display of delicate artworks? How do the toilets work? The design will be a more successful experience if such details support and reinforce (rather than contradict) the overall conceptual structure in a coherent way. Maintaining a consistent level of attention to quality when resolving such dilemmas creates a sense of unity and harmony that is evident to the visitor, even if subconsciously.

Information environments, too, have conceptual structures. Think back to the Chase bank website we visited in Chapter 3: it's divided into a "public" zone where you can access the bank's marketing materials and a "private" one where you can access your account details and functionality. To access the private zone, you must possess a "key," which is your username and password. As with a building, you can sketch out a diagram of how such an environment is structured.

Conceptual structure of the Kimbell Museum.

Instead of realizing these conceptual structures in marble, steel, and concrete, as built environments do, information environments implement them as words, phrases, icons, images, transitions, and sounds. Websites and apps don't have to resist the force of gravity, so they don't have load-bearing structures. Instead, we experience their semantic elements: the particular set of labels, icons, and so on that implement that environment's conceptual structure and create both distinctions and unity. Just as great care must go into the choice of materials in a building and the details of how they come together, the selection of words, metaphors, and images—and how they are rendered in the environment—are paramount to the success of an information environment.

How We Experience Structure

You may be getting the sense that we're talking about abstractions. However, conceptual structures have a tangible impact on the way you experience environments. Think of the physical place where you're reading these words. If you're inside a building, you will have experienced a series of transitions when going from one room to another. In one moment you're in a large public space (a lobby), the next you're in a smaller public space (an elevator), and then you're in a private office. As you go through these different areas, you form a mental model of how the place is structured—how you move from public to private spaces, for example. The elements that form the environment—its walls, doors, columns, windows, etc.—inform your understanding of its conceptual structure, and hence your mental model of the place.

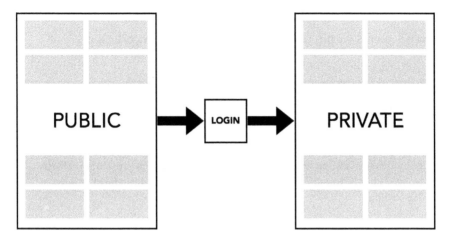

Conceptual structure of a bank's information environment.

Navigation and Labeling

Websites and apps obviously don't have columns, walls, or doors. Instead, you form mental models of these environments by interacting with their semantic structures: compositions of words, icons, images, and so on that you perceive in their navigation and labeling systems.

Navigation systems are the interactive elements that allow you (the user) to move from one part of the information environment to another. They take the form of sets of links or icons that you can click or tap on, or search systems that allow you to enter a text query and get back a set of results.

Labeling systems are predetermined, particular sets of words, phrases, and images you use to represent elements within an information environment consistently. Clear labeling systems give you an understanding of what is in the information environment and what you can do there. There is a clear relationship between labeling and navigation: the elements in navigation bars usually have labels. But labels also inform other structural elements within the environment, such as section headers.[3]

Microsoft Word on iOS has a primary navigation system based on tabs, which compartmentalize app functions. Note how tab labels are all verbs. Both of these features (tabs with verbs) work to help the user form a mental model of how to use the app.

Distinctions

The key to using language effectively in navigation and labeling systems is to create *distinctions* in the user's mind. The user won't be able to create a mental model of the place if all of its parts sound similar. To illustrate: if you see a link labeled "log in," you will immediately assume that the environment implements a distinction between public (logged out) and private (logged in) spaces. The difference between these is clear in your mind, and the use of the words *in* and *out* reinforces it.

The ways that you divide and label the parts that make up an information environment tell a lot about what that information environment *is for* and what you can expect to do there. For example, imagine if I show you an app with the following main navigation bar:

Studio   Classes   Poses   Schedule   Search

The semantic elements in navigation systems tell us much about what the information environment is meant to do.

---

3. For an in-depth discussion of labeling and navigation systems, see *Information Architecture: For the Web and Beyond* by Louis Rosenfeld, Peter Morville, and Jorge Arango (Sebastopol: O'Reilly Media, 2015).

What do you think such an app does? What is it meant to help you do? How are you expected to behave there? (Perhaps more of a stretch: What is its business model?)

You may have concluded (correctly) that this app is supposed to be a sort of digital yoga studio. How did you arrive at this understanding? The combination of labels ("Studio," "Classes," "Poses," "Schedule," and "Search") and icons formed distinctions in your mind that set each part of the environment apart from each other. (Classes are somehow different from poses.) Taken as a whole, they evoked a concept you already understood before you saw the navigation bar.

Note that this navigation system doesn't need to include the word "yoga" in it. By using terms from the technical vocabulary of the semantic environment of yoga ("studio," "classes," "poses," etc.), your mind starts thinking in that context. There is a sense of conceptual coherence and unity to these terms and icons (and even the choice of the color purple) that serve to reinforce the identity and purpose of the whole, even while presenting distinctions between its constituent parts.

Metaphors

The idea that the app functions as a yoga studio is an example of another concept that is central to the way you experience environments through their structure: metaphors. To remind you, a metaphor is a figure of speech that acts on a concept or context to which it does not apply literally. For example, you can't really "fall" in love any more than your laptop's screen is a "desktop."

Because so much of what information environments do is mirror and complement the things you do in the real world, metaphors are central to the way many of them work. There is nothing "natural" about the traditional computer desktop metaphor. "Files" and "folders" aren't stored in the computer as such; these terms help implement a metaphor that allows an alien world (that of bit sequences inside computer memory) to be understandable to humans by relating elements in that world to items in the physical world that people know and understand.

You can optimize information environments so they are more understandable by clarifying their underlying metaphors. We saw an example of this when Medium—an information environment designed to be a social network for writers and readers—changed the labeling and functionality of one of its key concepts. As is the case with many other social networks, Medium provides users with the ability to "like" individual posts. Before the change, users did this by tapping or clicking a "heart" button in each post. These buttons could be in one of two states: on or off. Tapping on the heart meant that the reader "liked" the post. The author could then have a sense of how well-liked (or not) the article was.

In the summer of 2017, Medium changed how this feature worked. Instead of an on/off "heart" button, users now saw a "clap" button. Clapping is how people are accustomed to rewarding good performances; the better the performance, the more they clap. Medium is leveraging this metaphor. Readers can tap the "clap" button multiple times to show how much they like a post. This new mechanism is more engaging for users since they can increase their level of approval as they scroll through an article and find more resonance. It's also better for authors since they receive more granular feedback. And, of course, it's also better for the system overall, since the concept of clapping itself reinforces the idea of Medium as a place where creative people can produce things that an audience can appreciate by echoing the way they do this in the "real" world.[4]

4. K. Zhu, "Show authors more 🖤 with 👏's," 3 Min *Read* (blog), Medium, August 10, 2017, https://blog.medium.com/show-authors-more-%EF%B8%8F-with-s-c1652279ba01

If you design software, you need to know about placemaking. Why? Because the websites and apps you design will create

♥   ⬆   🔖   280 recommends · 12 responses

If you design software, you need to know about placemaking. Why? Because the websites and apps you design will create

⬆   🔖   280 claps · 12 responses

Even though the functionality is similar, "recommending" something is not the same as "clapping" for it. The latter verb introduces a metaphor that is better suited to the purposes of the Medium information environment.

Thinking More Structurally

As you work on the design of the environment, you should ask yourself: What distinctions are necessary for this place to make sense to the people who will be using it? For example, if it's an online bank you're making, your customers will expect to be able to log in to a private part of the environment where they can see and interact with their account information.

Once you understand the way the environment should be divided, you must explore the relationships between these various parts. How easy will it be for people to move from one to the other? How will they know where they are? How will they know where they can go from there? Answering these questions in the terms of your customers requires that you delve into the language that makes the most sense to them, with the understanding that when you group individual words, you create contexts that change their meaning. The words *studio, classes,* and *poses* can have many meanings, but grouped together they evoke the context of yoga.

Thus, thinking structurally requires considering the parts that will make up the environment simultaneously with the whole those parts create. The two are interrelated in ways that make them impossible to separate. Aligning the distinctions and relationships between these parts with the needs and expectations of users can create an environment that empowers them. Doing so makes the environment more understandable, since the people using it can relate to the way it's organized, and it also makes it possible for them to focus their attention on completing the tasks they are there to do.

Conversely, structural distinctions can also be used to create environments that are at odds with the needs and expectations of the people who use them. This not only makes the environment more difficult to navigate, but it can also create environments that can be used to exploit users' attention against their interests.

The Test of Time

Back at the Hearst Theater, nine laser beams shoot from the floor in front of the stage toward the heavens in a fan-like arrangement. Jarre dramatically shoves his gloved hand into the path of one of the beams, interrupting it. A bright note immediately rings out from the theater's speakers.

The instrument he's playing—called a *laser harp*—would be incomprehensible to the Greek architects who designed the first prototypes of this building. So would all the other machines the audience had been experiencing throughout the evening: synthesizers, DJ decks, digital drum kits, amplifiers, and LED screens.

Our technologies have changed radically since the time of the ancient Greeks, as have our societies. In spite of this, we still find the amphitheater a useful place to hold public performances; Sophocles would probably have been baffled by Jarre's music, but he would have found the setting familiar. The conceptual structure of this environment is so well-attuned to its purpose and our sensory needs that it transcends cultural and technological changes that span many centuries.

While conceptual and practical structures are essential, they are not all that a great environment requires. For a place to be experienced as such, many other elements must come into play. For example, Jarre wouldn't be able to play his music at all if the theater had no electric outlets for him to plug in his equipment. The electrical system is only one of many that come together in a building to make it possible for us to use these environments to help us achieve our goals. You could say that a building is, in fact, a collection of systems working in concert to enable a place to happen. The same applies to information environments. In the next chapter, we'll see how they work.

Eventually, everything connects—people, ideas, objects. The quality of the connections is the key to quality per se.

—Charles Eames

8

Systems

A well-thought-out structure is a necessary component for a viable environment—but it's not sufficient. For you to be able to experience it, a building must be built; it must transcend the abstraction of its conceptual stage to become a thing in the world. Of course, the same goes for a website or an app. Making something as complex as a building or a software application requires that different systems work in concert toward a coherent experience.

Let's return to the Kimbell Art Museum. When you step into one of the Kimbell's galleries, you enter a space that takes care of various things you need. At the most basic level, the building provides a roof over your head. If it's sunny outside, you won't get baked, and if it's raining, you won't get wet. While the combination of materials and forms that make this roof possible is impressive, a building such as this goes much further than merely protecting you from the elements. For example, the air in the museum must be within a particular range of temperature and humidity for your comfort and the preservation of the art. Light washes down some surfaces and pools in others to provide consistent illumination and highlight particular objects and spaces. You shouldn't be distracted by unnecessary noise. Art displays are arranged in particular ways to allow you to move around them and observe them in the best light.

A gallery in the Kimbell Art Museum.

IMAGE: HTTPS://COMMONS.WIKIMEDIA.ORG/WIKI/FILE:KIMBELL_ART_MUSEUM_JANUARY_2017_2.JPG

Creating an environment such as this requires a thoughtful arrangement of various systems. They include the physical components holding up the roof, electrical components providing artificial illumination, HVAC (heating, ventilation, and air conditioning) components keeping the air comfortable, plumbing components allowing for water to come into and out of the building, telecommunications components, and more. The skillful combination of these systems adds up to a whole that is greater than the sum of its parts. Most environments you inhabit are complex systems composed of various subsystems, which you don't usually notice—when done right. If the Kimbell were too cold, or too noisy, or too dark, or too bright, it wouldn't work as a museum.[1]

Information environments, too, are complex systems composed of various subsystems. In the simplest of cases, they have data structures that store information, algorithms that manipulate it, and a user interface that makes this data accessible to everyone. The success of the whole—how well it allows you to accomplish your goals—depends on a skillful arrangement and coordination of various subsystems and components. To do a good job, the architect must consider them as an ensemble. She may consult with specialists to refine each of the subsystems, but the coherence of the whole relies on her understanding of how these subsystems fit together and contribute toward accomplishing the goals of the whole.

Consider a ride-hailing service such as Lyft or Uber. The Lyft application you install on your phone to request a ride from this type of service is only one component of a very complex system. Lyft is a marketplace that matches a demand ("I need to get somewhere") with an offer ("I have a car and can drive"). This marketplace calls for a variety of subsystems and components to work together. Besides the "passenger" app you

---

1. Gordon Pask: "Architects are first and foremost system designers who have been forced, over the last 100 years or so, to take an increasing interest in the organizational (i.e., nontangible) system properties of development, communication and control." G. Pask, "The Architectural Relevance of Cybernetics," *Architectural Design*, September issue No 7/6, 1969.

experience on your phone, there is also a "driver" app that drivers install on their phones. Both of these apps rely on a complex technology stack in smartphones and on the network. Elements of this stack include the internet, the Global Positioning System (GPS) to pinpoint your location, mapping services, user interface systems, financial systems to charge you and pay the driver, and more.

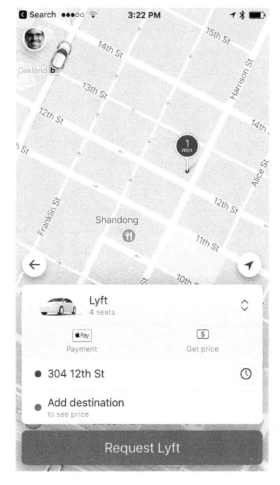

Consider all of the systems that make an application such as Lyft possible. An understanding of how these systems work in concert to create a marketplace is what makes the service work.

All of the components that make up the Lyft app ecosystem are incredibly complex systems on their own; just the mapping system would be a very high entry barrier if the company had to create it from scratch. The fact that such systems exist and that developers can easily recombine them into new systems is one of the advantages of working with modern information environments.

In any case, the key point is that just as physical environments such as buildings are systems composed of various subsystems, the same is true of information environments. Good design requires understanding how the whole works, and this, in turn, requires an understanding of what the parts and subsystems are and how they integrate and interact with each other in a way that creates a whole greater than the sum of the parts.

What Systems Are (and Why They Matter)

A system is a set of elements connected to each other in ways that allow them to form complex wholes toward a particular goal. While it may seem like a somewhat abstract subject, systems are everywhere; they are part of our day-to-day experience. But the fact that systems are pervasive doesn't mean they're obvious. Understanding systemic relationships calls for looking at things holistically (looking at the "big picture"). This is not easy. We usually want quick, easy answers to problems, so we break things down into smaller chunks that we can understand. Understanding complex wholes is unintuitive, difficult, and time-consuming. However, given how important systems are in the creation of information environments, it's important to understand them.

Let's look at an everyday example. You'll often hear people talk about wanting to lose or gain weight. (Mostly the former.) You'll see them modify their behavior in the pursuit of this goal. They'll eat less, eat different things, exercise more, or all of the above. They'll measure the effects of these actions by weighing themselves and taking note of the numbers to see how they vary over time. They will notice their clothes fitting more loosely, and will look at themselves in the mirror and notice changes to the shape of their body.

In doing these things, they have tapped into the power of a system. Diet and exercise are levers the person uses to influence this system. (Eating less will have one effect, eating more another.) Weight, the shape of their body, and the fit of their clothes are the feedback mechanisms that let the person know whether or not the levers produce the desired effect. Paying attention to what's going on allows these people to notice changes over time. If they vary how they eat and exercise, their weight and the fit of their clothes will vary.

This seems like a simple system. But as anyone who's tried to manage his or her weight knows, it's not easy to do in practice. That's because there's much more to the system than this simple model allows. For one thing, there are aspects of the system that are not obvious to the dieter. For example, varying the intake of food and exercise affects the body's metabolism, changing how it uses up energy. As a result, the effects of manipulating these control points are not linear. For another, the dieter's mind is also a component of the system—and the mind can do tricky things, like becoming frustrated at the lack of progress, or accepting exceptions "just this once," or going nuts at the office holiday party, or becoming disappointed by something someone else has said.

There are other components of this system that are even harder to see directly. For example, the pursuit of a particular body shape is a goal suggested to us by our culture, a broader system we participate in whether we want to or not. Also, some people's body chemistry is just different than others'; the "expected" behavior of the system will not work the same for these people. The stewardship of this system will vary depending on what the dieter's ultimate goal is. Their attitude and approach will be different if they're dieting in service to a healthier body than if they're dieting in service to a socially constructed image of what a "good" body looks like.

But enough of this weighty example. Let's focus now on your work. Any project you undertake will be subject to systems dynamics that have parallels to body weight management. You're expected to perform to certain standards as you strive toward particular project goals. Perhaps you've committed to deliver some artifact by a particular date, and there are consequences tied to your ability to meet this goal. Working toward the goal may require some struggle on your part; perhaps you need to stay up late for a few nights in a row or put off going on a vacation. Incentives drive you to make these sacrifices; perhaps there's a bonus if you deliver on time (carrot), or you may lose your job if you fail (stick). You'll recall these as remunerative and coercive incentives, respectively.

Whatever the case, you're participating in this system. Your effectiveness will at least depend on

- Whether you understand the goals you're driving toward (e.g., delivering a particular thing to a particular standard by a particular date)

- Whether you understand what you're being measured on (e.g., is the date more important than the quality of the work?)

- Whether the appropriate feedback mechanisms are in place (e.g., regular check-ins to validate that things are going in the right direction) and everyone understands them

You'll notice that all three of these conditions call for better communications. For this system to work effectively, you and the person who has requested the work need to be "on the same page"—i.e., speak the same language. Most of the problems I've encountered in projects have been due to communications breakdowns. This is something you must proactively work at, because in many cases these systems lack good communication mechanisms. And, of course, your mind is also a component of the system. The same mind tricks that can trip you up when managing your weight can affect your performance at work. This is aggravated by the fact that most people work in groups, and groups bring with them interpersonal dynamics and politics, which add a great deal of noise and complexity.

Let's look at a few critical systems concepts by examining a complex system you're familiar with: your body.

A System Has Goals

Systems are not just random collections of parts: the parts work together toward achieving one or more goals. For example, your body consists of various organs (your liver, brain, lungs, etc.) and subsystems (your digestive system, your circulatory system, etc.) working in concert to make a more complex whole: *you*. These organs and subsystems are all working to support their own goals. For example, the digestive system converts food into energy your body can use. And, of course, the broader system that is your body is working toward a goal as well: staying alive.

A System Has Resources That Ebb and Flow

Systems require resources to operate. Some of these resources are stored in the system as stock for future use, while others flow through the system as needed. For example, you can't go very long without breathing because the body doesn't have mechanisms to keep oxygen in reserve. On the other hand, your body also requires energy, which it acquires mainly from the food you eat. A few days without eating won't kill you, since the body stores some of this energy as fat. You can map the flow of energy in the body as it undergoes these various transformations.

A System Has Feedback Mechanisms

Systems have the means to monitor the state of its components and adjust its functions accordingly. Your body lets you know when it needs more energy (sleepy! hungry!), when it needs to cool off, and more. If you feel thirsty, you drink, which keeps an essential resource—water—in balance within your body. You'd be in deep trouble if these feedback mechanisms failed.

A System Has a Boundary

The collection of parts that make up a system is not limitless: there's a clear difference between the world inside the system and the world outside. For example, your body makes energy from food, a resource that comes from outside the system. You put it into the system through your mouth, and your digestive system breaks it down so the necessary chemicals can enter your system. In the case of your body, the boundary between the system and the outside world is palpable: it's the surface of your skin.[2]

System Resources

I've been illustrating the characteristics of systems by looking at your body. But our objective here is to create more effective environments, so let's return to the building example. As you saw earlier, a building such as the Kimbell is a complex system composed of subsystems. An art museum requires certain resources to accomplish its goals. To start with, erecting the building requires energy, time, materials, and effort by humans and machinery. Then there are ongoing operational requirements: the building takes in energy for lighting and air conditioning, potable water, products for sale at the gift shop, the time and attention of the people who operate it, and so on. The system outputs heat, sewage, refuse, and particular experiences, among other things.

Traditionally, building designers have had to think systemically about the way the many components and subsystems that comprise the building work together. However, many architects focus instead on the way this impacts the relationship between the forms of buildings and the spaces they create. Recently, some have also started to consider how

---

2. Where we place the boundary of a system is dependent on what we're interested in studying. If we're looking at the individual organism, the skin may be a useful boundary. But if we're interested in how the organism procures its food, we need to expand that boundary to include the organism's environment.

resources flow into and out of buildings, how they manage stocks of these resources, and the impact these systems and resource flows have on the broader environments in which they operate.

For example, the Adam Joseph Lewis Center for Environmental Studies at Oberlin College in Ohio is a building whose primary design goal was to be a sustainable system. Among other things, the Lewis Center treats its own wastewater and produces the electric energy it requires to operate. (It's so efficient at doing so that it eventually became a net energy exporter, producing 30% more energy than it needs.) The site and orientation of the building keep temperatures comfortable while maximizing natural lighting, and its materials were sourced responsibly.[3]

In short, the team that designed the Lewis Center brought to the project an understanding of systems that went much deeper into the way buildings function over time than traditional building designs. Given that the Lewis Center houses an environmental studies program, the whole-systems approach is particularly apt for its goals. However, such an approach would also benefit all buildings. According to the U.S. Green Building Council, buildings produce 39% of carbon dioxide emissions in the United States.[4] Since most of these emissions come from burning fossil fuels to meet electricity needs, buildings that produce their energy as the Lewis Center does would help reduce CO_2 emissions.

We have a good understanding of how resources flow into and out of built environments and how to make them more efficient; doing so is mostly a matter of having the right incentives in place. However, when it comes to information environments, we don't yet have a good understanding of the resources required to operate them sustainably. Most efforts thus far in this direction have focused on making our computers more energy-efficient, with devices getting smaller and batteries lasting longer every year. If you've used smartphones for some time, you may have realized that some apps use up your battery faster than

3. https://www.bdcnetwork.com/living-machine
4. http://www.eesi.org/files/climate.pdf

others. The software that runs your smartphone can be made more energy-efficient by improving its underlying software and hardware and through various application-programming techniques. The same applies to the computers in data centers that host many of our information environments.

However, one resource we haven't considered thus far is the attention of the people using these systems. Given that our time is limited, I think our attention is our ultimate nonrenewable resource. Whether we're creating an information environment that helps people diagnose and treat diseases or one that helps them gossip with their friends, we're going to be using up part of their allotted time on earth. Are they getting their "time's worth" for engaging with this system? Is the environment encouraging them to bring other people along, as is the case with many social networks, and if so, will their time be well-spent? Much as we've learned to not squander precious natural resources like clean water and air, we need to learn not to waste our attention. Those of us who design apps and websites have a responsibility to ensure that the systems we create have a net positive "attention footprint" on the people who use them. To do so, we have to think about them as systems that take attention as an input.

How to Think More Systemically

One of the most challenging aspects of designing information environments is seeing beyond their user interfaces to the conceptual structures that underlie them, and how elements in those structures interact with each other. During the design process, it's especially tempting to want to jump to sketch out screen layouts; UI (user interface) is relatable and appealing, and therefore easier for stakeholders to discuss. However, it's essential that user interface elements—particularly those in which structure is most naturally expressed, such as navigation and labeling systems—manifest a coherent structural direction. So how do you explore and express these abstract ideas in ways that you and others can understand? You do so by using concept models and maps.

Before architects start to design the form of a building, they often explore the possible relationships of its spaces using *bubble diagrams*, simple sketches that eschew such critical details as materials and physical structures in favor of more abstract representations. These diagrams allow architects to run quick what-if scenarios to solve high-level project goals, such as meeting functional requirements, establishing the right balance between private and public spaces, and enabling efficient circulation. The building's "look and feel" can be resolved later, once these deeper issues have been resolved.

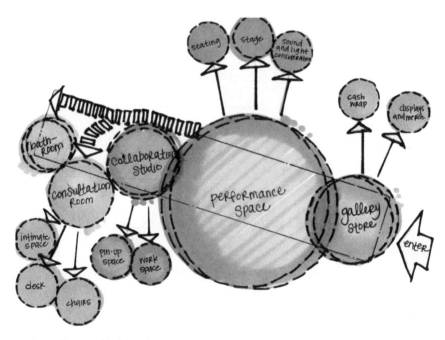

An architectural bubble diagram.

SOURCE: LILYGLOVER.BLOGSPOT.COM/2009/10/HATCHERY-BUBBLE-DIAGRAM.HTML

In designing information environments, the equivalent of a bubble diagram is the conceptual model. These are drawings that articulate the concepts the users of a system will be interacting with and the relationships between them. They allow you to explore the system's component parts, the relationships between them, the resources they make use of, its feedback mechanisms, the goals it works toward, and the boundaries between it and other parts of the world.

As with bubble diagrams, designers can use conceptual models to quickly evaluate high-level what-if scenarios without getting mired in discussions of the "look and feel" of the finished product. Conceptual models are essential to the design of effective information environments, and an excellent tool to help you start thinking more systemically about the work you do.[5]

Technically, creating a conceptual model is not difficult: You can sketch one on a whiteboard or a large piece of paper. The challenge lies in thinking about what you're working on in the abstract and recognizing that in many cases the people using the system will be different from you. It's also important to remember that these people have goals they want to accomplish using the system and that these goals may be misaligned with the business objectives of the organization that is commissioning the system. Bringing user and business goals into alignment (at best) or symbiosis (at least) is essential.

Whenever I'm mapping out one of these systems, I find the following questions useful:

- What domain of human activity will this information environment be hosting? (Healthcare? Financial services? Interpersonal relationships? Or whatever.)

- Who will be using this environment?

- How much do they understand about this domain before engaging with this environment?

- What do they expect to find there?

5. For a deeper discussion of this subject, see Jeff Johnson and Austin Henderson, *Conceptual Models: Core to Good Design* (San Rafael: Morgan & Claypool, 2011).

- What do they call those things?
- What are the most natural divisions for the parts of the environment?
- Do those individual components need to be systems themselves?
- How do people expect those components to relate to each other?
- What unexpected relationships between elements must the environment accommodate?
- What goals do users expect to accomplish in this environment?
- What business goals must this environment serve? In what time frame?
- Are the business and user goals in alignment?
- Are the business and user time perspectives aligned?
- Where are the boundaries of the environment?

As Simple as Possible, but Not Simpler[6]

Let's delve into this final question. One of the biggest challenges that designers of information environments face is defining the boundaries of the system. In this sense, building designers have it easy: The boundary of a construction project is defined by the lot on which the structure will be erected and other limitations imposed by physics. For example, the dimensions of the average car will dictate how many of them will fit in a particular lot, and no amount of haranguing will increase the density of automobiles afforded by a flat piece of land of limited size.

Because designers of information environments are unshackled from the constraints of the physical world, they must place artificial limits on the extent of the system lest the system becomes unwieldy and unpractical. But they need to do so carefully: every system has

6. An aphorism attributed to Albert Einstein. See https://quoteinvestigator.com/2011/05/13/einstein-simple/

a minimum number of elements and relationships that it requires to serve its goals. Cutting out more than is needed will result in a system that can't do what it was designed for.

I always like to keep in mind Gall's law, which states:

> A complex system that works is invariably found to have evolved from a simple system that worked. A complex system designed from scratch never works and cannot be patched up to make it work. You have to start over with a simple working system. [7]

I often bring this law to the attention of clients who have wide-ranging ambitions for their information environments. During the design process, we often clarify a vision for what the environment will be once it's fully operational. Those visions are based on a series of assumptions that need to be tested. The best way to do this is to begin with the most minimal set of elements possible, and evolve the system from there over time as they have a better understanding of how users interact in and with the environment. The final vision is not meant to be fully realized on the day the system launches; the vision merely hints at the direction in which they want the system to evolve.

This point brings up an important fact we have not touched on thus far in our discussion: Systems are never static; they change over time to adapt and respond to varying external and internal conditions. For example, a new competitor may have entered the market, changing your app's value proposition. Or perhaps your organization is looking to serve a new segment of the market. Or maybe a new technology has become available that makes your product more effective. Whatever the case, environments are always in transition: if they are not being maintained and changed, they slide toward irrelevance. In the next chapter, we'll look at how to design information environments that can successfully stand the test of time.

---

7. John Gall, *Systemantics: How Systems Really Work and How They Fail* (The General Systemantics Press, Ann Arbor, Michigan, 1975).

When we build, let us think that we build forever.

—John Ruskin

9

Sustainability

"This design will still be new and fresh 50 years from now, we think.... What we have is magnificent." These were the words of A. L. Scott, President of the Kimbell Art Museum's board of directors, after the building's opening.[1] Time has proven Mr. Scott right. Since it opened in 1972, Kahn's building has come to be recognized as an excellent example of museum design and a modern classic.

Given the expense and effort that goes into making buildings, it's natural that they should aspire to longevity. Some, like the Kimbell, achieve it. But like all systems, buildings are always changing. On the one hand, you have the forces of nature relentlessly wearing down physical materials; left unmaintained,

---

1. https://fashionpluslifestyle.wordpress.com/2013/11/14/the-kimbell-art-museum-unveils-newest-acquisition-a-luminous-pavilion-by-renowned-architect-renzo-piano-november-27-2013/

buildings deteriorate quickly. On the other, you have changing conditions—both internal and external—that change how you understand and use the building. For example, new technologies may appear that make a particular building's use irrelevant. Or perhaps the character of the neighborhood around the building changes, forcing new approaches. Or tastes shift and what seemed fresh and hip one year is old-fashioned and irrelevant after a decade. Or the organization that commissioned the building moves on, and another one—with a different set of needs—occupies the premises.[2] Whatever the case, as stable as built environments may appear to be, they don't stay the same for long.

This malleability is even more true of information environments, which don't have to contend with the constraints of physical materials. Deploying a change to the navigation structure of a mobile app requires design, development, and testing efforts—but it doesn't require bulldozers and building permits. Information environments also exist within a context—that of information technology—which is evolving very fast. For example, iOS 7 introduced a completely new visual design to the iPhone's operating system. From one day to the next, the feeling of the entire information environment changed, and perfectly functional applications that didn't immediately implement the new style suddenly looked old and out-of-place. This change was experienced by millions of people literally overnight.

So if environments are to stand the test of time, they must be able to accommodate change. However, they can't change thoughtlessly, lest they fail to serve their intended functions. People must feel like they know the place when they visit; an environment that is changing in radical ways from one day to the next would be difficult if not impossible to use. Thus, the environments that best serve their goals over the long term, as the Kimbell does, strike a balance between flexibility and

2. I've taught classes at a former bus depot in the San Francisco campus of the California College of Arts.

stability. They provide coherence and understandability while evolving gracefully in response to changing conditions. In other words, their structures and systems must be resilient.

Resilience

I define resilience as the ability of any system—including environments such as the ones we've been discussing—to respond and adapt to change without compromising its primary purpose or its integrity. Change can be incremental, such as erosion caused by the elements or the gradual reduction of a critical resource. It can also be sudden. For example, I work in an old building in Oakland, California, which is in a seismically active zone. This building has been retrofitted so it can continue functioning safely as a building (its primary purpose) after being subjected to the forces of an earthquake (sudden change).

In seismically active regions like the San Francisco Bay Area, there are companies that specialize in making older buildings more resilient to earthquakes.

This concern with the resilience of the environment is not new. One of the earliest building codes in the world appears in the Code of Hammurabi, which is over 3,700 years old. It states:

> 229 If a builder builds a house for someone, and does not construct it properly, and the house which he built falls in and kills its owner, then that builder shall be put to death.

> 230 If it kills the son of the owner, the son of that builder shall be put to death.

> 231 If it kills a slave of the owner, then he shall pay, slave for slave, to the owner of the house.

> 232 If it ruins goods, he shall make compensation for all that has been ruined, and inasmuch as he did not construct properly this house which he built and it fell, he shall re-erect the house from his own means.

> 233 If a builder builds a house for someone, even though he has not yet completed it; if then the walls seem toppling, the builder must make the walls solid from his own means.

The intent here is not to specify how structures are meant to be resilient, but rather to incentivize builders to do so. Using incentives such as these, architects have long had "skin in the game" of keeping our environments useful and safe over time.

Information environments, too, are subject to changes. Some of these are internal, such as the launch of a new product or service. Others are external, such as the appearance of a major new platform or interaction mechanism. For example, after the introduction of the iPhone in 2007, most people access information environments using small touchscreen-based devices. Many information environments designed before 2007 had to adapt to the constraints and possibilities of this platform to remain relevant.

This ability to adapt and change in response to changing conditions is essential if an environment is to continue serving its purpose over time. However, it shouldn't be taken for granted. Several factors will determine how resilient an environment can be, including its size, complexity, malleability, and degree of dependence on other systems. Beyond that, much rides on the people responsible for managing the environment:

- The team must have a healthy *attitude* toward change. They should embrace change as natural and expectable and have the willingness to respond by altering the environment accordingly.

- The team must have *awareness* of what is happening within and without the environment. They can only respond appropriately to change if they can perceive it.

- It may be that the team sees what needs changing, but lacks the resources or political support to respond. Thus, the team must be *empowered* to respond to changes.

- The team must have a *clear vision* of the purpose and essential character of the whole and how people use it. Understanding the whole is important if the team is to respond without compromising the environment.

- The *design* of the environment must accommodate change gracefully. Some do this better than others; much depends on the environment's structural configuration.

In order to be resilient, environments need to be *sustainable*. You can think of sustainability as creating the conditions necessary for a system to meet the needs of its present stakeholders without compromising the needs of its future stakeholders. In the case of the physical environment, the primary goal should be to ensure that it can usefully host your activities in the long term. When dealing with an information environment, your goal should be to make sure that it can host meaningful interactions in the long term. To do this, it must sustain:

- **Itself:** The environment should be able to generate enough resources to support its continued existence.

- **Its purpose:** The environment should generate these resources without compromising the reason(s) why it exists.

- **Its social context:** The environment should achieve its purpose(s) without compromising the societies that host it.

These goals mirror the goals of sustainable development formulated during the 2005 World Summit on Social Development: the *economic, social,* and *environmental* "pillars," or fundamental aspects of the system. Let's see how they map to your work.

Economic Sustainability

Creating and maintaining an information environment requires resources. These resources include labor to design, build, test, and manage software; servers to host it; energy to power them; infrastructure to deal with logistics; and more. The system should be able to generate enough value to produce the resources necessary to ensure its continuing existence. This seems like an obvious statement, perhaps one not even worth mentioning. However, some information environments have gained tremendous social importance without having a business model that points to their continuing viability.[3]

Social Sustainability

Information environments exist within a broader social construct. For them to remain viable in the long term, society as a whole must continue to be viable as well. By *viable,* I mean the society must continue

---

3. The most notorious of these is Twitter, which has become an important means for the U.S. Government to convey policy decisions, even though it hasn't yet demonstrated a path to sustainable growth.

to work well for the people who participate in it without compromising itself or the environments it exists within. The social fabric must encourage cooperation between diverse people toward common goals.

Again, this seems like an obvious thing to say. However, many information environments depend on business models that, while viable from the economic perspective, may be socially unsustainable. For example, advertising-based business models can be problematic, since advertising drives us toward more consumption and does so by targeting us as members of ever-narrower demographic segments. Given the challenges we face as a society, we should be striving instead to be more mindful of our consumption and more focused on the things that unite us.

Environmental Sustainability

Information environments create communication ecosystems that can either sustain or harm our societies' long-term prospects. We need to consider their impact on these ecosystems. As you saw in Chapter 2, communication happens in semantic environments that have parallels to physical environments. The goal of these semantic environments is to *convey meaning*. Like the physical environment, semantic environments can become polluted, making them incapable of achieving these objectives. In semantic environments, pollution happens when the language, rules, and purposes of one particular semantic environment (e.g., science) start to become blurred with those of another (e.g., religion).

For example, after the 2016 U.S. election, there was much talk about the problem of "fake news" in social networks. What this means is that a particular semantic environment (social media, which we're using to inform our worldview) is becoming polluted with material from another semantic environment (outright propaganda, or in some cases, satire). The effect, overall, is to erode the meaning of the word "news," making certain types of conversations more difficult.

As you saw in Chapter 4, disinformation has been around for a long time. However, the pervasiveness of information technologies, and the ease with which we move between semantic environments, make today's information environments particularly prone to spreading disinformation. It behooves the designers and operators of these environments to understand how they can become polluted, and work to ensure that the transmission of meaning can happen as "cleanly" as possible.

Designing Environments That Accommodate Change

As you saw in Chapter 7, information environments have underlying semantic structures. Similar to the load-bearing structures in buildings, these semantic structures change more slowly than other aspects of the information environment. So if you want to create information environments that successfully convey meaning and maintain their integrity even as they evolve, you must carefully design their conceptual structures and the semantic structures that implement them. This is especially important because these semantic and conceptual structures tend to be longer lived than other aspects of the environment.

I'd understood this idea conceptually, but it became very tangible when I was working on the fourth edition of *Information Architecture: For the Web and Beyond*. One of my tasks in that project was updating the examples in the book, which required that I revisit many of the websites featured in previous editions. One of those websites—FedEx.com—had a very different visual presentation in the mid-2000s (when the third edition of the book was written) than it did 10 years later. However, when I started examining the site's semantic structures, it struck me as to how little they had changed over that decade. I've seen this in my work as well. It's not uncommon for organizations to overhaul their websites' and apps' "look and feel" while leaving their primary categorization and language schemes mostly untouched.

Buildings, too, have some elements that change more slowly than others. Stewart Brand's book *How Buildings Learn: What Happens After*

They're Built popularized the *shearing layers* model, which was originally proposed by the architect Frank Duffy. The idea is that buildings are composed of layers that change at different rates. These layers are (from slowest to fastest): site, structure, skin, services, space plan, and stuff. "Site"—the ground upon which the building rests—changes very slowly, at a geological pace. "Stuff," on the other hand, refers to the things we put inside buildings, such as furniture and appliances, which can easily be moved and therefore change much faster. As buildings adapt to evolve, the form they take is affected by the differences in the malleability of the various layers.[4]

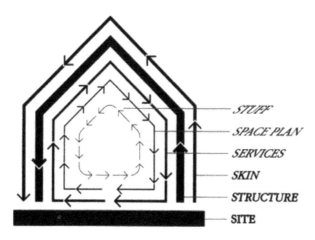

Brand's "shearing layers" model (1994).
STEWART BRAND, *HOW BUILDINGS LEARN: WHAT HAPPENS AFTER THEY'RE BUILT*

Brand subsequently extended the shearing layers model to explain how civilizations change. This broader model is also composed of six layers (again, from slowest to fastest): nature, culture, governance, infrastructure, commerce, and fashion. Brand explains that because fashion (and art) change so quickly, this is where civilization experiments with new

---

4. Stewart Brand, *How Buildings Learn: What Happens After They're Built* (New York: Penguin Books, 1994).

ideas and ways of being. Worthwhile ideas are assimilated into the underlying layers, where they become more permanent parts of the civilization. As Brand explains,

> The combination of *fast and slow* components makes the system resilient, along with the way the differently paced parts affect each other. Fast learns, slow remembers. Fast proposes, slow disposes. Fast is discontinuous, slow is continuous. Fast and small instructs slow and big by accrued innovation and occasional revolution. Slow and big controls small and fast by constraint and constancy. Fast gets all our attention, slow has all the power.
>
> All durable dynamic systems have this sort of structure; it is what makes them adaptable and robust.[5]

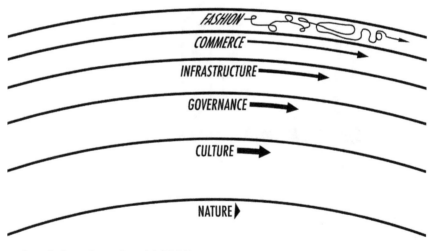

Brand's "pace layers" model (1999).
IMAGE: HTTP://LONGNOW.ORG/ABOUT/

5. Stewart Brand, *The Clock of the Long Now: Time and Responsibility* (New York: Basic Books, 1999).

This is a very useful insight. It helps us understand how the unevenly changing parts of a system can help make it stronger as it evolves.

As I've been thinking about how to make information environments more conceptually sound and resilient, I've started mapping my work to a pace-layer model. These are the layers I've come up with, from slowest to fastest:

- **Purpose:** Why the organization, team, or product exists. This is not a goal since it can never be achieved; it's an aspiration that the system is always working toward.

- **Strategy:** How the organization aspires to do things differently to strive toward its purpose; how it's going to compete.

- **Governance:** How the organization shapes itself to implement its strategy. What are the rules and means of engagement, including the organization's internal hierarchy?

- **Structure:** The relationships between particular semantic elements that will inform end products and services.

- **Form:** The user interfaces that people use to interact with the organization's products and services. This layer is where the structure is articulated as artifacts that humans can experience.

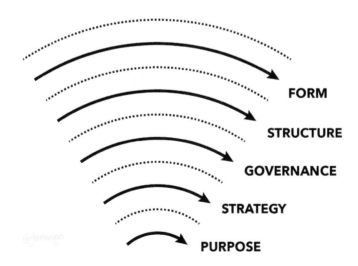

I separate form from structure for two reasons:

- As I mentioned earlier, the structure that informs these products and services changes more slowly than the user interfaces that are built upon it.

- You experience the things you design through apps, websites, social media, and a variety of other touch points. For the sake of coherence, an organization's various user-facing artifacts should share a common semantic structure.[6]

Many designers spend a disproportionate amount of time focused on the form layer. This emphasis is understandable. The user interface (UI) is where the "new and shiny" action is happening. It's also much easier to discuss UI artifacts since the environment's structure is more abstract.

Note that the first three layers aren't commonly thought to be the domain of designers at all; the last two are where designers are usually brought in. However, to be successful, designers should be conversant in all of the layers and move effortlessly among them. The governance layer in particular has an important impact on the design of information environments. This is where business rules and processes live. It's essential that designers understand how the business works as a system and how that impacts their work.

In any case, understanding which layer you're acting on at any given time is key to being effective as change agents since all the layers require different approaches. Designers need to acknowledge that structural decisions are going to be a part of their information environments for longer than their user interfaces. If you expect these environments to last, you need to pay careful attention to their structural underpinnings.

---

6. In *Pervasive Information Architecture*, Resmini and Rosati refer to this as *consistency*.

Assessing Resilience

Often when you're in the midst of designing or operating an information environment, you can lose sight of the bigger picture it works within. It's especially difficult for teams internal to the organization to perceive the opportunities and shortcomings inherent in their evolving internal and external contexts.

Asking these questions can help you take a measure of the state of the environment and the degree to which it can evolve gracefully:

- Are there aspects of the environment that used to work and are no longer working as intended? How do you know?

- Do you have mechanisms to address those aspects of the environment that are no longer working? (How do you know when things are going wrong? What can you do about it?)

- Are there new requirements that the current environment doesn't serve?

- Has the context outside the environment changed?

- Has the context the environment creates changed?

- How do you know that these changes have happened? (Do you trust these sensing mechanisms?)

- Is everyone in the team clear on the original vision that drove the design of the environment?

- Is that original vision still relevant?

- Is this still a single environment? Or does it need to be broken up to pursue separate visions or strategies?

- What is the environment in service of?

What Is It in Service Of?

This final question is of particular importance and worth pondering at length. The organization may have a stated vision of the sort of change it wants to effect in the world, but that is not necessarily what it or its information environments are really in service to. More obviously, there may be differences between what an organization claims to be working toward, and the signals it sends as it goes about implementing its strategy. As always, actions speak louder than words.

Consider the case of Microsoft. From humble beginnings in the 1970s, the company rose to lead its field in the 1990s on the strength of founder Bill Gates's vision of having "a computer on every desk and in every home." Microsoft's strategic moves and product-level decisions were in service to this vision, and it came close to achieving its vision with its Windows operating system. However, by the 2010s, it was clear that something had gone wrong. Most people didn't do their computing tasks on *desktops* anymore but instead were using small mobile devices that didn't run Windows. Unable to see the changing ecosystem it was now operating in (CEO Steve Ballmer famously scoffed at the iPhone after its introduction), the company doubled down on its strategy by refusing to give non-Windows customers access to its key Office software suite. (While there was a version of Office available for Apple's Macintosh platform, Office for Mac was vastly inferior to its Windows counterpart.)

While still outwardly adhering to another vision, at this time Microsoft was actually in service to preserving the hegemony of the Windows operating system at all costs. And cost it did: market share (and mind share) kept eroding as customers sought alternatives to Microsoft products in their iPads and other mobile devices. In February 2014, Ballmer stepped down and a new CEO—Satya Nadella—took over. Nadella brought a new vision to the company, one centered on enabling customer productivity on a variety of devices and contexts, including those of competitors like Apple and Google. Shortly after Nadella took over, Microsoft released versions of Office for iPhones, iPads, and Android devices. Microsoft's products may not regain the level of ubiquity they had during the company's heyday, but at least now they're available in the most relevant personal computing platforms of today.

These vision and strategy decisions had a substantial impact on the design of Microsoft's information environments. Office is an information environment that people experience in a variety of devices and contexts. In an internal email to colleagues, Nadella explained how he sees the company's products in a mobile-first world: "Our worldview for mobile-first is not just about the mobility of devices; it's centered on the mobility of *experiences* that, in turn, are *orchestrated by the cloud.*" (Emphasis mine.) This statement hints at a direction for the structure and front-end design of Microsoft's customer-facing information environments. It calls for coherence across platforms in the front end, and consistency in the back end. This change in vision and strategy is manifest in the design of the products, with Microsoft touting the consistent experience that users of products such as OneNote have when using diverse devices.[7]

So how do you determine what an information environment—and the organization that created and maintains it—is ultimately in service of? For commercial enterprises, the most reliable way I've found is to "follow the money": to examine the means by which the organization procures the resources necessary to operate and profit from its information environments. For example, I believe Apple exists in service of creating excellent, desirable electronic devices that enhance people's lives. I believe this because the company's primary source of revenue is

---

7. https://blogs.office.com/en-us/2017/05/18/note-taking-made-easier-for-everyone-redesigning-onenote/

selling such devices to consumers. If Apple profited primarily from sell-ing their customers' attention (by selling advertising, for example), it would focus on market penetration instead of quality. When I use one of Apple's information environments, such as *Apple Music*, I trust that my information and activities within the environment aren't employed in ways that compromise my privacy.[8]

Stewardship

As we've seen, information environments are complex systems. As with all systems, they must adapt to changing conditions within them and in their broader contexts. In the best cases, this process allows these envi-ronments to continue serving their purposes for a long time. However, this best case is not a given. For an environment to evolve gracefully without compromising its purpose, it must be managed by stewards who understand the vision that led to the creation of the place and who have a clear understanding of the conditions inside and outside of the environment.

Let's return one last time to the Kimbell Art Museum. By the late 1980s, the Kimbell's art collection had outgrown Kahn's original design. In 1989, museum director Ted Pillsbury announced plans to expand the building by adding two wings to its north and south ends. This proposal triggered an outcry from the architectural commu-nity, which at this point regarded Kahn's building as a masterpiece. Eventually, the museum's leadership dropped plans to alter the build-ing. Instead, they commissioned another celebrated architect—Renzo Piano, who worked with Kahn at one point—to create another building

---

8. Apple has set out to differentiate itself from its competitors by taking a rigorous stance on issues of customer privacy and security. https://www.nytimes.com/2016/02/21/technology/apple-sees-value-in-privacy-vow.html

across from the original Kahn structure. This strategy honors Kahn's original vision,[9] and the new building acknowledges its celebrated predecessor by maintaining a similar structure and size, without mimicking it.[10] The decision to expand by creating an entirely new environment is never an easy one. It takes conscious leadership to determine when it makes more sense to expand the existing place and when to break off and create a new one.

We've thus far been discussing buildings and architecture as a model for the structuring and organization of information environments. However, as with all models, this one falls short. Given the fact that physics doesn't encumber them, information environments can be much more dynamic and organic than buildings. Because of this, we will now examine stewardship of information environments more closely by adopting as a model another type of human-designed environment: the garden.

9. According to a group of prominent architects, Kahn argued that any new addition to the Kimbell should be in the form of a new building located across the lawn from the original building. http://www.nytimes.com/1989/12/24/arts/l-kimbell-museum-in-praise-of-the-status-quo-384789.html
10. Piano has stated that the new building is "close enough for a conversation, not too close and not too far away." https://www.kimbellart.org/architecture/piano-pavilion

The ability to self-organize is the strongest form of system resilience. A system that can evolve can survive almost any change, by changing itself.

—Donella Meadows

CHAPTER

10

Gardening

Thus far we've seen how websites and apps create contexts that influence the ways we understand the world and act on it. I've suggested they have taken on some roles that buildings have played over time. However, built environments differ from information environments in many ways. Not least among these differences is the rate at which they change.

Buildings and cities change over their lifetimes, but they do so slowly. The California College of the Arts, where I teach, is housed in a building that at one time was used as a bus depot. The structure wasn't designed to serve as host for university-level courses, yet it's evolved to serve that role through deliberate design interventions. This sort of significant change doesn't often happen to a building, but it does happen.

Information environments, on the other hand, must accommodate more frequent change. I'm currently working on a project for a large software maker that will change their website's entire navigation structures. When this new navigation system launches, the structure of this information environment will change considerably, and it will happen overnight. This is not the first such change to this environment; in fact, this particular website went through a major redesign not long ago.

Changing conditions in the market and inside the organization, and changing expectations from users, are always exerting pressure on information environments. Compared to the effort required to transform a bus depot, changing the structure and form of a website is relatively easy. Thus, the nature of information environments is to be in a constant state of flux.

Because the structure and form of an information environment will always be changing, its designers must create conditions that allow it to evolve and adapt without having its integrity and objectives compromised. Here our architectural analogy starts to break down: built structures are perceived by many people as being much too rigid to accommodate the rates of change needed, and the architectural design process much too definitive. For this, we must look elsewhere.

Designing the Thing That Designs the Thing

The musician and record producer Brian Eno has been influential through his work and collaborations with David Bowie, U2, Coldplay, Talking Heads, and others. In a 2011 speech at the Serpentine Gallery in London, Eno spoke of discovering a new way of making music in the mid-1960s. Previously, he'd assumed music was first imagined by a composer, written down as a score, and handed over to musicians who would perform it as written. Eno contrasts this traditional approach of creating music with that of composers such as John Cage, Terry Riley, and Steven

Reich, who work differently. Instead of writing a musical score upfront, these composers write rules that generate musical pieces.

Performances of music "written" through such an approach are never the same twice, but still maintain an identifiable character. For example, two separate performances of Riley's "In C" are never identical, but you can still tell you're listening to "In C." The "composition" consists of rules that define boundaries for the piece, allowing for variations within a prescribed range of possibilities. In other words, this way of composing substitutes the upfront definition of a particular order with the definition of a *system that generates order.* Eno describes the subject of his Serpentine lecture as

> [...] the shift from "architect" to "gardener," where "architect" stands for "someone who carries a full picture of the work before it is made," to "gardener" standing for "someone who plants seeds and waits to see exactly what will come up."[1]

The "seeds" in this analogy are an initial set of sounds and algorithms that interact and evolve to generate new pieces. In contrast to composers of more traditional music, composers of this "generative music" don't design static sonic landscapes; they design systems that produce dynamic sonic landscapes.

As we move to create more complex information environments that continuously evolve, it becomes impossible for a single designer or team of designers to consider all possible requirements or explore all possible permutations up front. A new approach is called for, and Eno's generative composition technique hints at a way for us to work toward a specific design objective without prescribing structures or forms up front. Gardens aren't "wild" nature, nor are generative music pieces random noise: like buildings, they both have structure and intentionality. They're designed artifacts, but their exact "final" form is unknown (and unknowable) to their creators.

---

1. https://www.edge.org/conversation/brian_eno-composers-as-gardeners

If we want to create information environments that serve our needs and stand the test of time, we must move to a more generative approach that allows structures to evolve continuously. Taking a page from Eno's book, I call such environments *generative*, in contrast to the types of environments with prescribed structures that information architects have traditionally produced. Let's examine the differences between them.

Prescribed Structures

Environments with prescribed (or top-down) structures are defined up front by the design team, and aren't expected to change too much. These structures result from a design process that takes as its input the requirements known to the design team when they're designing the environment. Consider your bank's website: the categories in the main navigation bar were defined by a designer or team of designers working for the bank in consultation with other people in the organization, and were coded into the bank's website by developers.

Citibank.com's "logged in" primary navigation structure. This structure doesn't change frequently.

When changes to such a structure are required, the team will engage in a redesign project that sometimes brings significant (and possibly traumatic) changes to the environment. These redesigns aren't considered part of the regular day-to-day operation of the system, but rather they are exceptional occurrences that must be carefully planned and budgeted for.

There's an obvious downside to this approach. Because the bulk of the work in designing a prescribed structure happens up front, it's critical that the team get the requirements right. This is a tall order. For one thing, many requirements aren't obvious until the team has a prototype version of the system that can be stress tested under real conditions.

For another, the context in which the design process is happening continues to evolve during the project, and requirements that may have been critically important during the definition phase may be less so by the time the system is ready to launch. Designers and stakeholders can't predict with certainty the context in which their products will exist a year or two in the future. We'd be better served by creating up-front structures that support continuous change and adaptation.

Generative Environments

Instead of thinking of design interventions as exceptional events in the lifespan of the environment, we should plan for its continuous evolution. This requires us to give up our current ideas of control and ownership over the exact form and structure of our information environments.

Environments built from computer code are dynamic, fluid, and interactive. They all have the potential to change themselves dynamically to address changing needs and context. Whenever you visit the Amazon.com home page, you will see slightly different offers based on what the company wants to promote and your past shopping and browsing behaviors. A good part of this content selection is algorithmic—in other words, chosen in real time by computer code instead of human hands.

Every aspect of an information environment—from its structure to its content—is implemented in code, so every aspect is available for programmatic manipulation. This includes navigation elements, the sequence in which content is presented, and the content itself. Think of the Facebook news feed: contrary to what its name implies, the sequence in which items appear on that list is not chronological, but determined by an algorithm that examines your patterns of behavior in the system to show you content in the order which it thinks will engage you most. (You can tweak this algorithm by marking some posts as "liked" and hiding others; this "trains" the system on the things you like and dislike.) The ability to change dynamically based on user activity and awareness of context makes it possible for you to create environments whose structures aren't fixed upfront, but that evolve.

Generative environments can serve your needs better than environments with prescribed structures because they don't require clairvoyance. The structure and form of environments that support emergence can adapt over time to better serve their purpose, and keep evolving as the context keeps changing. Generative environments also tap on our collective knowledge, enriching the environment and making it more resilient through diversity. Because their users also participate in their creation, generative environments can also be better aligned with their users' interests.

Characteristics of Generative Environments

Generative environments require a new approach to design, with different principles than those we use to design prescriptively. For one thing, generative environments call for us to attend as much to the dynamics of the system and the people who will be creating it as we do to structure and form.

The foremost example of a generative information environment that supports emergence is Wikipedia, which has grown from a modest effort driven by a small group of individuals to become one of the most popular and useful websites in the world. Wikipedia started from the premise that its creators didn't have all the answers; what they had was a vision of what the environment should be.

Instead of designing its structures from the top-down, Wikipedia's designer's created a system that enabled other people to design and build the place from the bottom-up. As such, Wikipedia serves as a good illustration of the characteristics of generative information environments. Let's examine some of the most important of these characteristics.

Unfinished

Generative information environments are always unfinished. As with all complex systems, they aren't static, but ever changing in response to changing conditions inside and outside them. You could say they're in a *continuous state of becoming* what they aspire to be—even if they never truly achieve it.

A snapshot of a system at any moment will quickly be out of date as the system keeps evolving. During the time it takes for your web browser to download and display a Wikipedia article, many changes have occurred to the site in the background. New articles have been added; existing articles have changed (including, perhaps, even the article you're downloading); new links have been created between existing articles; and so on. Your article is a small part of a great evolving system; if you were to download the entire Wikipedia to your computer, your local copy would be obsolete by the time it finished downloading.

Collaborative

Generative information environments make it possible for people to work with each other to contribute to the system. People can easily "onboard" to become actors in the system, interact with each other (and with nonhuman agents) in the environment, and see the effects of their collective actions. Note this doesn't mean the system is a free-for-all: there must be rules and boundaries that keep things from devolving into anarchy and incoherence.

Wikipedia provides mechanisms for editors to communicate with each other as they go about making changes to articles. These discussions happen in public, and anyone can view them by clicking on the "Talk" tab at the top of any article.

The "Talk" tab for the Wikipedia entry on Emergence.

Diverse

Generative information environments enable diversity within a stable framework. This makes it possible for actors with varying perspectives to contribute, thus making the whole richer and more resilient.

One of the reasons why Wikipedia has grown to become as useful as it has is because it makes it possible for specialists in all fields to contribute their knowledge to the whole. No one person can know in-depth about every subject, but there are individuals who are deeply knowledgeable about their particular field of interest. If you entice enough such individuals to contribute their knowledge to the system, eventually you have a resource that has both depth and breadth.[2]

---

2. It's worth noting that while Wikipedia takes submissions from all sorts of people, they've been called out on the perceived lack of gender diversity in their editorial team. The managers of generative information environments must aim to proactively increase diversity in their systems, and Wikipedia could do better in this regard. See https://jezebel.com/5978883/wikipedias-editors-are-87-percent-male-because-citations-are-stored-in-the-penis

Robust

Generative information environments are robust: they can change without compromising the integrity of the whole. Robustness calls for a balance between flexibility and stability. One way to accomplish this is by making it possible for actors in the system to revert potentially catastrophic changes. Wikipedia can delegate power and autonomy to new actors in the system because the environment is *forgiving*: changes to individual articles can be rolled back easily to earlier states. Thus, actions from a rogue actor don't threaten the system overall. (You can see the evolution of a Wikipedia article by clicking on its "View History" tab.)

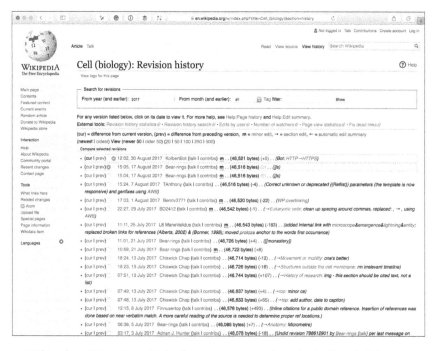

Wikipedia's "View History" tab allows you to see a chronological list of all the changes to an individual article.

Another way to encourage robustness is by providing different degrees of flexibility at different levels of the environment's structure. Contributors to Wikipedia have a great degree of freedom in what they can change in articles; for example, they can choose to correct a small spelling mistake or wipe out entire pages. Contributors can also create new article pages. However, they can't change global structural elements such as the site menu bars, header, and footer. Changes to these elements must be done by stewards of the system—actors who have developed a trustworthy reputation over time.

Too-rigid structures are brittle, while too-pliable ones lack the strength to give the whole direction. Aim for a mix of the two. Think of a tree: the leaves and branches are pliable while the trunk is relatively solid; this allows the tree to remain upright even while swayed by the wind.

Reputable

Knowing who is allowed to make particular types of changes to the environment requires tracking the reputation of actors within it. If you know nothing else about contributors to the system, seeing a history of their past actions can help you make decisions about how trustworthy they can be. (This is why credit scores exist.)

Wikipedia stores all changes to the environment and where they came from, regardless of whether the individuals who made those changes are known actors within the system. This changelog is visible to everyone who uses Wikipedia, so users can trace back the history of changes to an article and who made them. "Anonymous" Wikipedia editors only have agency at the level of individual articles. Other actions in the system—such as deleting articles altogether—are not open to everyone, but require some degree of reputation within the environment.

Thus, Wikipedia implements a hierarchy of authority: new users can undertake some low-level actions, while higher-level actions are only available to users who have been vetted by various means, including their history of interactions within the system.

Transparent

Generative information environments provide mechanisms for actors to understand what is going on at any given time, both with the elements they are responsible for and with the whole. When they sense a change that requires their attention, they can then respond appropriately.

For example, the editor of an article in Wikipedia cares about changes made to that article. Articles are editable by many people at the same time, so changes can happen faster than anyone can manage by simply reloading article pages in a web browser. Contributors to Wikipedia have created tools with names like *Vandal Fighter* and *WikipediaVision* to allow editors to monitor changes to articles in real time so they can respond appropriately.

Recursive

A generative information environment allows actors to do more than add and modify the content in it. It also provides mechanisms for those actors to modify the structure and operation of the environment itself. Because they are made of code, the rules that define these environments are not materially different than the content those rules produce. So they can be reinterpreted and adapted by the actors in the environment as easily as they can change and add to its content.

This doesn't mean that the system is a free-for-all. If anyone could do anything they wanted to the rules and structure of an environment at any time, it wouldn't last long. Generative information environments strike a balance between adaptability and incoherence by making it possible for the system to adapt to changing conditions while remaining oriented toward its goals.

Wikipedia provides mechanisms for the rules that guide it to change. These rules can't be modified arbitrarily by any one individual in the system, but are subject to a consensus among a group of trusted actors.

Users can become trusted actors over time by demonstrating responsible behavior and commitment to the whole.

Led

Environments that support emergence are led by people who are invested in the long-term success of the system. They deeply understand the vision that animates the environment and the principles that support the community as they move toward that vision.

Such leaders see themselves as stewards of the place. They demonstrate a strong sense of responsibility for the long-term viability of the system and understand this requires a delicate balance between bottom-up decision making and calling the shots from the top. They don't try to micromanage the place; instead, they focus on creating the right culture and ensuring that things are moving along in accordance with the vision.

While Wikipedia allows for a great deal of flexibility in the way it runs, it isn't a democracy. In a 2005 TED talk, Jimmy Wales—the founder and leader of Wikipedia—described his role thus:

> […] if you look at most of the major free software projects, they have one single person in charge who everyone agrees is the benevolent dictator. Well, I don't like the term "benevolent dictator," and I don't think that it's my job or my role in the world of ideas to be the dictator of the future of all human knowledge compiled by the world. It just isn't appropriate. But there is a need still for a certain amount of monarchy… sometimes we have to make a decision, and we don't want to get bogged down too heavily in the formal decision-making processes.[3]

---

3. https://www.ted.com/talks/jimmy_wales_on_the_birth_of_wikipedia

Cohesive

One of the most important tasks that fall on the environment's leaders is creating a sense of cohesion among actors in it so they all know what they're working toward and how they should go about it. There are various ways to do this.

To start with, it's essential that actors in the environment have a clear understanding of what the place is. Is it a forum for discussing books? A bank? An online academy? Whichever it is, actors must understand that they're working on it together. This requires that they have a clear vision of what the place aspires to be.

A strong vision clearly articulated helps actors align in a particular direction, motivating them to implement changes that benefit the whole rather than the individual parts under their direct control. This is especially important for environments that allow for fundamental changes to the environment since these can alter the nature of the place.

Wikipedia's vision is unambiguous:

> Wikipedia is an encyclopedia. It combines many features of general and specialized encyclopedias, almanacs, and gazetteers. Wikipedia is not a soapbox, an advertising platform, a vanity press, an experiment in anarchy or democracy, an indiscriminate collection of information, or a web directory. It is not a dictionary, a newspaper, or a collection of source documents, although some of its fellow Wikimedia projects are.[4]

This description is from the first of Wikipedia's "Five Pillars," the fundamental principles of the system. They are:

- Wikipedia is an encyclopedia.
- Wikipedia is written from a neutral point of view.

---

4. https://en.wikipedia.org/wiki/Wikipedia:Five_pillars

- Wikipedia is free content that anyone can use, edit, and distribute.

- Wikipedia's editors treat each other with respect and civility.

- Wikipedia has no firm rules.

This last pillar is worth delving into in more detail:

> Wikipedia has policies and guidelines, but they are not carved in stone; their content and interpretation can evolve over time. The principles and spirit matter more than literal wording, and sometimes improving Wikipedia requires making exceptions. Be bold but not reckless in updating articles. And do not agonize over making mistakes: every past version of a page is saved, so mistakes can be easily corrected.[5]

Instead of providing fixed rules—which could become brittle and constraining over time as conditions change—Wikipedia provides general *principles*. The system is successful because it trusts its contributors to do the right thing. What the right thing *is* is up to their interpretation of the principles that define the system, given the vision that propels them.

The vision, principles, particular approach, and path to reputation-building create in participants the sense that they are part of a special group of people on a mission. This gives them a sense of belonging, which, in turn, contributes to their having a stake in the success of the system. People on a mission who feel they are special will look after each other and their joint work. Wikipedia encourages this sense of identity by giving contributors to the environment a name: *Wikipedians*.

Wikipedia's vision is rooted in *purpose*: making information about the world more easily accessible to people. As you saw in the previous chapter, a clear purpose informs all that follows.

---

5. https://en.wikipedia.org/wiki/Wikipedia:Five_pillars

Designing for the Only Constant

We've been looking at Wikipedia as an example of an environment that supports emergence, and you may have been thinking something along the lines of, "An open encyclopedia is very different from a bank or a hospital. How would this model work in environments that require more top-down control?"

This is an important question. There are many reasons to want more control. Some fields, such as financial services and healthcare, are highly regulated, and the information environments that serve them have fiduciary and legal liabilities for the organizations that operate them. It's hard to imagine how these places could be opened to co-creation by a broad community.

However, even these environments must change over time. The market changes; technology changes; people move on to other roles; with time, they die. The fact of the matter is, *all* environments support emergence—some just do it more slowly than others. As you saw in the previous chapter, even massive, solid things like buildings and cities change over time. The only thing you can know with certainty about any designed environment is that the conditions and actors that led to its creation will change. The bank's website may not be open to contributions to anyone in the world like Wikipedia is. Still, its designers must consider who will keep the place going after its initial stakeholders are no longer around.

If our environments—buildings, towns, websites, apps—are to have a long useful life—and serve our needs in the long term—we must plan for their ongoing evolution under different stewardship and in different conditions. In other words, *all environments should be designed for emergence*. Ultimately, the difference isn't between prescribed and generative structures: it's between environments designed by people who understand systems and change—and those designed by people who don't.

We are called to be architects of the future, not its victims.

—Buckminster Fuller

Conclusion

If you ask people what they think about when they think about design, many will tell you about things they either like or dislike about a particular product. You'll hear about their iPhone, their car, a building, a chair, a book, a poster. It's always about a *thing*—a form that exists in the world. This shouldn't surprise us. We can relate to forms. We see them, touch them, hold them, get into and out of them. They're "real"; we tacitly understand where we stand in relation to them.

But forms are not the only product of design. Things don't exist in a vacuum; they *always* address—and alter—a broader context. The coffee mug next to your computer is a response to a context that includes your biological need to ingest liquids, the mechanics of your body, a culture that has taught you to prefer coffee hot, and so on. A chair hints at a particular course of action, and its dimensions and materials respond to physical, economic, and social constraints. A room with a video camera in it changes your behavior. (This is why public places visibly announce their presence.)

Context is not as easy to perceive as forms are. You can't touch context in the same way you can touch an iPhone or a coffee cup or a chair. Instead, you experience the effects of acting within a context when the forms that enable it alter your understanding and behavior.

Forms can be explicitly designed to create particular contexts. Consider Albert Speer's design for the Nazi party rally grounds (*Reichsparteitagsgelände*) outside of Nuremberg:

PHOTO VIA WIKIMEDIA, HTTPS://COMMONS.WIKIMEDIA.ORG/WIKI/FILE:NAZI_PARTY_RALLY_GROUNDS_(1934).JPG

The forms that made up this place were subservient to the context they were designed to create: a place where individuality was discouraged, and social hierarchies and rules were made concrete—literally. This place was in service to reinforcing a broader context—that of the Third Reich—which produced the conditions that called for the creation of the Reichsparteitagsgelände to begin with.

So context births forms and forms alter context in a cycle of constant adjustment. The Reichsparteitagsgelände (along with many other intentionally designed forms) was created by—and helped create—a context that encouraged and enabled unspeakable atrocities. When the war was over, the forms that had enabled this context had to be eradicated.

Cognitive scientist George Lakoff writes about the influence of framing on our behavior. Frames, he explains, are mental structures that shape the way we see the world. As a result, they shape the goals we seek, the plans we make, the way we act, and what counts as a good or bad outcome for our actions.

You experience framing in the language you use when you talk about things. When Lakoff introduced the concept to his students in Berkeley, he asked an impossible task of them: *Don't think of an elephant.* (This is also the name of Lakoff's 2004 book about the power of framing and language in politics.[1]) The exercise is impossible because the instant the word "elephant" is uttered, the student has no choice but to think about one. (You're doing so now.) The frames you use when you talk about things have an important influence on how you think about them. Critically, you can change the frames you use.

When you talk about digital things as products, services, publications, or interactions, you're using frames that impose on them metaphors that influence how you design and use them. *Product* suggests a commodity that is meant to be bought, sold, and used; a good to be consumed. *Services* and *interactions* are transactional, ephemeral: an interaction happens; a service is rendered, and conditions are changed, the actors move on. A *publication* implies a distinction between an author and a reader, a subtle yet critical hierarchical distinction.

It's not that any of these frames are right or wrong; they can all be used to describe aspects of digital things. Rather, you should be asking yourself: what visions of the world are these framings supporting? How do they affect what you understand to be good or bad outcomes? You have a choice in how you talk about things, and how you talk about them influences how you do them.

1. George Lakoff, *The ALL NEW Don't Think of an Elephant!: Know Your Values and Frame the Debate* (White River Junction: Chelsea Green Publishing, 2014).

In the spring of 2017, I was at a reception for a group of interaction design students who were graduating from a prestigious college in the San Francisco Bay Area. Knowing that I was focused on information architecture, a friend who taught there told me that when talking with new students, he realized most of them either didn't know what IA was or thought it was something that no longer mattered.

The Bay Area is home to the most influential information environments in the world. Some of these students will go on to work at companies like Facebook, Google, Twitter, and Uber. Their work will create contexts that will affect how millions of people understand their world and make decisions that will impact the lives of many millions more.

Using *architecture* as the framing for the design of digital things would open the door for them to think with a longer-term focus. Buildings are not meant to be consumed; they are designed to last. They have cultural import. They can naturally host the provision of services and interactions, the commercial exchange of products, and the production and consumption of publications. Architecture has been the broader frame in which we've done these things for centuries.

Thinking through the architectural structures and systems that underpin our websites and apps is a prerequisite to creating contexts that support our needs in the long-term. And it's not enough that we focus on technical excellence; we must also strive for ethical excellence. When it comes to digital products and services, this means working toward information systems that extend and augment our personal and social abilities. It means abandoning notions of creating environments with top-down, prescribed structures, and moving toward design that aspires to create generative environments that stand the test of time.

Generative information environments help us make decisions, which further our best interests as individuals, as organizations, and as societies. They allow us to have access to the information we need, when we need it, and help us discover new connections between concepts. They expose us to a broad spectrum of thought and help us form our perspectives on the issues that matter.

Generative information environments respect and value our attention. They don't aim to "engage" but to elucidate. They acknowledge that we're not just consumers or (in Silicon Valley parlance) "users," but multifaceted and complex. They allow us to do what we came for and then set us free to go our ways without cajoling us.

Generative information environments create more value than they capture.[2] The business models that support them reward all participants in the environment, without making them subject to another. These models are transparent; they allow participants to understand where the resources that allow the environment to operate come from.

Generative information environments are resilient. They generate the resources needed to allow them to perform their duties sustainably over the long term. Doing so calls for thinking beyond technology to questions of governance and organizational structure.

Generative information environments do not compromise the viability of society as a whole. They avoid imposing distinctions that further divisions and prejudices among social groups. They make it possible for people to participate in civil discussions without feeling threatened, persecuted, or spied on.

Digital information technologies can be the most powerful force for good the world has ever seen. Never before have so many people accessed so much knowledge so cheaply, nor connected with each other so effortlessly. But positive use is not inherent in the technology itself;

---

2. I took this phrase from Tim O'Reilly, who's used it as one of the guiding principles for his company O'Reilly Media. Tim O'Reilly, *Tim O'Reilly in a Nutshell* (Sebastopol: O'Reilly Media, 2004).

it must be consciously sought and designed for. When used carelessly or maliciously, these same technologies can trap our minds in opinion bubbles, foster addictions, and create contexts that lead to personal and societal ruin.

Generative information environments will not emerge organically. They require intentional design—they require *architecture*. And if they are to remain generative over the long term, they also require *stewardship*. This calls for leadership, vision, and the courage to think deeply, broadly, and long term.

Index

A

advertising

  buying attention online, 55–56, 58–59

  funding information environments, 62–63

  in newspapers, 53–55

  social unsustainability of, 141

Aesop's fable of boy who cried wolf, 52–53, 56

affordances, in physical environments, 23–26

agency, as factor that drives incentives, 37–38

agents, in environment, 22–23

Alexander, Christopher, 91

aligned incentives, 42–44

Amazon.com

  artificial intelligence of algorithms, 71–72

  generative environment of, 157

  shopping in information, 15

Anderson, Laurie, 64

anonymity, in identity factor that drives incentives, 39–40

Apple

  design of corporate headquarters, 97

  in service of, 148–150

architectural bubble diagrams, 130–131

architectural history, One Montgomery Street, San Francisco, 1–3

architecture, 81–97. *See also* buildings

  design of environments, 87–89, 90

  design principles, 89, 91–92

  design process, 84–87

  as framing for digital things, 172

  generative information environments and, 174

  Great Workroom at Johnson company, 81–82, 97

  information architecture (IA), 93–96

  mental models, 92–93

  product-oriented design, 83, 97

The Architecture of Information (Dade-Robertson), 19

artificial intelligence (AI), 70–71

attention

  buying and selling, 52–56, 58–59

  generative information environments and, 173

  harvesting by media, 55

  meaning of, and why it matters, 48–49

  as nonrenewable resource, 129

augmented reality (AR), 68–70

B

Ballmer, Steve, 148

banks. *See also* Chase bank
  aligned incentives of, 42–43
  architectural history, 1–3
  incentives in, 33–36
  prescribed structures, 156–157

BART metro system, xiv

Bates, Marcia, 46

"being" somewhere, 51–52

blockchain, 78

body weight management,
  as system, 123–124

boundaries, in systems, 127, 132–133

Brand, Stewart, 40, 142–144

brand safety, 58

bubble diagrams, 130–131

Buchanan, Richard, xi

building codes, 138

buildings. *See also* architecture
  design of, 81–82, 84, 87, 97
  entrances to, 11–12
  shearing layers model, 143
  sustainability of, 135–136

C

California College of the Arts, 153

carbon dioxide emissions, 128

cave art, 4

Centro GAM, Santiago, Chile, 6

Chartres Cathedral, 9

Chase bank, 42–43, 109

chat application, as information
  technology, 13–14

Churchill, Winston, 20

Citibank.com, 156

cities, affordances and
  signifiers in, 24–26

*Cities: Comparison of Form and
  Scale* (Wurman), 94

civilizations, and shearing
  layers model, 143–144

Code of Hammurabi, 138

coercive incentives, 36–37, 40, 125

coffeehouses, 59–61

coherence, as design principle, 91

cohesive, generative environment
  as, 165–166

collaborative, generative
  environment as, 159

community, third place as, 5

companies, structure of, 102–103

computers
  as information technology, 13–14
  TRS-80 Model I, 79
  wearable, 75

conceptual models, in systems, 130–132

conceptual structure, 107,
  109–110, 116–117

context, 21–31
  agents in environment, 22–23
  forms and, 169–170
  physical environments and
    affordances, 23–26
  semantic environments, 26–31

conversational user interfaces, 74

Cooper, Alan, xvii

credit cards, trust delegated to
    third parties, 77–78

D

Dade-Robertson, Martin, 19

Day, Benjamin, 53–54

design
    of environments, 81–82,
        87–89, 90, 97
    of generative environments, 154–156
    process, 84–87
    within systems, x–xiii

design principles
    coherence, 91
    ergonomics, 89
    fit, 91
    function, 89
    quality, 91
    resilience, 91–92
    understandability, 89

Digital Ground (McCullough), 18–19

disinformation, 141–142

disintermediating trust, 77–78

distinctions in structure, 112–113

diverse, generative environment as, 160

Doblin, Jay, x–xi

Dubberly, Hugh, xiii

Duffy, Frank, 143

E

Eames, Charles, 118

earthquakes, resilience to, 137

economic incentives, 57–63

economic sustainability, 140

education
    learning in information, 16
    learning online, xv

elevator safety brake technology, 65–66

emergence, designing for, 167

employment. See also office
    environments
    working in information, 15–16

encyclopedias, 17–18, 165, 167

engagement, 47–63
    attention, buying and selling, 52–56
    attention, importance of, 48–49
    being somewhere, 51–52
    economic incentives, 57–63
    foraging for information, 50–51

Eno, Brian, 154–156

environmental sustainability, 141–142

environments, 1–19
    architectural history, 1–3
    design of, 81–82, 87–89, 90, 97
    designing to accommodate
        change, 142–146
    information, 9–12
    information environments, 12–19.
        See also information environments
    physical environments, 3–8

ergonomics, as design principle, 89

Esslinger, Hartmut, x

F

Facebook
  design of corporate headquarters, 97
  generative environment of, 157–158
  influence in presidential election, xv
  misaligned incentives, 45
  news feed artificial intelligence, 72–73
  selling attention to advertisers, 58–59
"fake news," 141
FedEx, 142
feedback mechanisms, as system concept, 126
filter bubble, 73
fit, as design principle, 91
foraging for information, 50–51
Forlizzi, Jodi, xi
forms, and context, 169–170
Foster, Norman, 97
frames, 171
Fuller, Buckminster, 168
function, as design principle, 89
future, technology of, 79

G

Gall's law, 133
Gates, Bill, 148
Gehry, Frank, 97
generative environments, 153–167
  change, rates of, 153–154
  continuous evolution of, 157–158
  design of, 154–156
  designing for emergence, 167
  prescribed structures, 156–157
generative environments, characteristics of, 158–166
  cohesive, 165–166
  collaborative, 159
  diverse, 160
  led, 164
  recursive, 163–164
  reputable, 162
  robust, 161–162
  transparent, 163
  unfinished, 159
Gibson, Eleanor, 24
Gibson, JJ, 24
goals, as system concept, 126
Goldberger, Paul, 81–82
Google, design of corporate headquarters, 97
graphical user interfaces (GUIs), 73
The Great Good Place (Oldenburg), 5
Great Moon Hoax of 1835, 54, 57
Great Retail Apocalypse of 2017, xv, 15
Great Workroom at Johnson company, 81–82, 97
greedy environments, 62

H

Hammurabi, Code of, 138
Harris, Tristan, 61
Hearst Greek Theater, 99–100, 102, 116–117

Heatherwick, Thomas, 97

Hellenistic theater at Ephesus, 100

Hephaestus, 70

Her (film), 70

Herschel, John, 54

hierarchies, as structure, 102–103

High Court at Chandigarh, 12

Hinton, Andrew, 19, 22

holacracy, 103

How Buildings Learn: What Happens After They're Built (Brand), 142–143

human-computer interaction (HCI), 93

IBM research, 93

icons in software navigation system, as semantic element, 29

identity, as factor that drives incentives, 39–40

incentives, 33–45

    agency, as factor that drives, 37–38

    aligned, 42–44

    in banks, 33–36

    identity, as factor that drives, 39–40

    misaligned, 44–45, 57

    power imbalances, as factor that drives, 38–39

    transparency, as factor that drives, 40–41

    types: remunerative, social, and coercive, 36–37

information, defined, 9–12

Information Architecture: Blueprints for the Web (Wodtke), 96

Information Architecture: For the Web and Beyond (Rosenfeld, Morville, and Arango), 96, 142

Information Architecture for the World Wide Web (Rosenfeld & Morville), 95

information architecture (IA), 93–96

information environments, 12–19

    described, 2–3

    learning in information, 16

    placemaking with information, 17–19

    shopping in information, 15

    socializing in information, 16–17

    working in information, 15–16

information foraging theory, 50–51

information sciences, 95

information technologies, 12–14, 67, 136

Ingels, Bjarke, 97

interactions, and metaphorical frames, 171

interactive information environments, 55–56

internet, as information technology, 13–14. *See also* websites

Internet of Things (IoT), 75–76

iPhone, 136, 138, 149

Ito, Joi, xii

Jarre, Jean-Michel, 99, 116–117

Jobs, Steve, x, 37

Johnson, Herbert, 81, 97

Johnson, Mark L., 98

Johnson company, 81–82, 97

K

Kahn, Louis, 94, 106, 135, 150

Kimbell Art Museum, Fort Worth, Texas, 106–109, 119–121, 135, 150–151

L

labeling systems, in structure, 111–112

Lakoff, George, 171

language

    as information technology, 12–13

    smart assistants' use of, 73–74

laser harp, 116

lattices, as structure, 103

learning in information, 16

led, as characteristic of generative environment, 164

Lewis Center for Environmental Studies, Oberlin College, 128

libraries as physical environments, 21–23, 101

light bulb, Philips Hue, 75–76

log in link, 112

Lyft ride-hailing service, 121–123

M

Maeda, John, xii

Mandiberg, Michael, 17–18

matrixed structures, 103

McCullough, Malcolm, 18–19

Meadows, Donella, 152

Medium, information environment, 114–115

mental models of environment, 92–93

metaphors in structure, 113–115

Microsoft, in service of, 148–149

Microsoft Word

    labeling system, 112

    semantic environment of, 28–29

misaligned incentives, 44–45, 57

MOOCs (massive open online courses), 16

Morville, Peter, 95

music composition, 154–155

N

Nadella, Satya, 148–149

National Mall, Washington, D.C., 7–8

navigation bar in software, as semantic environment, 28–30

navigation systems, in structure, 111–112

Nazi party rally grounds, 169–170

Nelson's Column, London, 90

networks, as structure, 103

newspapers, selling advertising, 53–55

nightclubs as physical environments, 21–23

Norman, Don, 25

O

office environments

    Great Workroom at Johnson company, 81–82, 97

    place in cube farm office, 5

Oldenburg, Ray, 5, 16

One Montgomery Street, San Francisco, 1–3

online advertising, 55

online retailing, 15

optimal foraging theory, 50

Ortega y Gassett, José, 52

Otis, Elisha, 65–66

P

pace layers model, 144–145

Palo Alto Research Center
(PARC), Xerox, 50, 93

Pask, Gordon, xii

pattern-matching, 106

patterns of order, 104

Pervasive Information Architecture
(Resmini & Rosati), 19, 95, 146

Philips Hue light bulb, 75–76

physical environments, 3–8

context and affordances, 23–26

places and roles, 4–5

physical level of places, 7

physical structure, 99–102, 107–109

Piano, Renzo, 150

Pillsbury, Ted, 150

placemaking with information, 17–19

places, 4–8

physical and symbolic levels of, 7

polar bear book, 95, 96

Polk, Willis, 1

Porter, Michael, xi

POS (point of sale) terminals, 77–78

Postman, Neil, 26

power imbalances, as factor that
drives incentives, 38–39

prescribed structures, 156–157

private spaces *vs.* public spaces, 109–111

Procter & Gamble, 58

products

digital environments as, 83, 97

and metaphorical frames, 171

public spaces *vs.* private spaces, 109–111

publication, and metaphorical frames, 171

Pygmalion, 70

Q

quality, as design principle, 91

R

realities, of information
technologies, 68–70

recursive, generative environment
as, 163–164

religion and science as semantic
environments, 26–27

remunerative incentives, 36–37, 39, 125

reputable, generative environment as, 162

resilience, 137–142

assessing, 147

defined, 137

as design principle, 91–92

economic sustainability, 140

environmental
sustainability, 141–142

factors of, in environment, 139

generative information
environments and, 173

social sustainability, 140–141

Resmini, Andrea, 19, 96, 146

resources of systems, 126, 127–129

retail shopping, 15

Riley, Terry, 154–155

robustness, of generative
environment, 161–162

Rosati, Luca, 19, 96, 146

Rosenfeld, Louis, 95

Ruskin, John, 134

S

Schlossberg, Edwin, 14

Schultz, Howard, 59

science and religion as semantic
environments, 26–27

Scott, A. L., 135

Scott Brown, Denise, 80

semantic environments, 26–31

    pollution by disinformation, 141–142

    science and religion as, 26–27

    software as, 28–30

services, and metaphorical frames, 171

shearing layers model, 143–144

shopping in information, 15

signifiers, 24–26

signs on lawns, 10–11

Sinclair, Upton, 32

Slack, 82–83

smart assistants, 73–74

smart devices, 75–76

smartphones

    design with information
architecture, 96

    system resources of, 128–129

social incentives, 36–37, 39–40

social networks, socializing in
information, 16–17

social sustainability, 140–141

socializing in information, 16–17

software design

    as placemaking, 19

    as semantic environment, 28–30

spaces, public *vs.* private, 109–111

Speer, Albert, 169

Starbucks, 59–61

stewardship, 150–151, 174

structure, 99–117

    conceptual, 107, 109–110, 116–117

    distinctions, 112–113

    experiencing, 110–115

    meaning of, and why it
matters, 102–110

    metaphors, 113–115

    navigation and labeling, 111–112

    physical, 99–102, 107–109

    prescribed, 156–157

    thinking structurally, 115–116

Sullivan, Louis, x, 89

The Sun newspaper, 54, 57

sustainability, 135–151

    of buildings, 135–136

    described, 139–140

    designing environments that
accommodate change, 142–146

    economic sustainability, 140

    environmental
sustainability, 141–142

    resilience, 137–142

resilience, assessing, 147
in service of, 148–150
social sustainability, 140–141
stewardship, 150–151
symbolic level of places, 7
systems, 119–133
boundaries of, 132–133
concepts of complex
systems, 126–127
meaning of, and why
they matter, 123–125
resources of, 127–129
from simple to complex, 133
thinking systemically, 129–132

T

technology, 65–79. *See also*
information technologies
connected things, 75–76
disintermediating trust, 77–78
elevator safety brake, 65–66
history and the future, 79
machines that decide for us, 70–73
machines that speak our
language, 73–74
reality reframed, 68–70
TED talks, 94, 164
Temple of Artemis at Ephesus, 105
thermostats, smart, 76
third parties, trust delegated to, 77–78
third places, 5, 16, 59
Tokyo transit map, 94–95

touchscreens, machines that
speak our language, 73
transparency
as factor that drives
incentives, 40–41
in generative environment, 163
trust, 77–78
Tufekci, Zeynep, 57–58
Twitter, xv, 38, 39–40, 41
2001: A Space Odyssey, 70

U

understandability, as design principle, 89
Understanding Context (Hinton), 19, 22
unfinished, generative
environment as, 159
user interface (UI), in system
design, 129, 146

V

value, generative information
environments and, 173
viability
generative information
environments and, 173
of social sustainability, 140–141
virtual reality (VR), 68–70
Vitruvius, x

W

Wales, Jimmy, 18, 164
wayfinding, 18, 96
wearable computers, 75

websites

    design of, 95–96, 156–157

    placemaking with information, 18

    as semantic environments, 29–31, 142

WELL, online community, 40

Wells Fargo bank, 1–3

Widener Library, Harvard University, 101

Wikipedia, 17–18, 158–167

    changelog, 162

    Five Pillars principles, 165–166

    Talk tab, 159–160

    View History tab, 161

Wikipedians, 166

Winter Palace, St. Petersburg, Russia, 88

Wodtke, Christina, 96

Wood, John Turtle, 105

working in information, 15–16

Wright, Frank Lloyd, 81

Wurman, Richard Saul, 94–95

Y

YOYOW (you own your own words), 40

Z

Zappos, 103

Acknowledgments

Even though my name is on the cover, this book is a product of a rich, evolving conversation within a global community of information architecture practitioners and academics. This conversation has spanned many years and environments (physical and otherwise), so it's impossible for me to individually acknowledge everyone who contributed to my thinking on the subject. But to all who actively participated in discussions about the discipline of information architecture over the years: *thank you.*

I called upon four members of this community for technical feedback: Marsha Haverty, Andrew Hinton, Dan Klyn, and Andrea Resmini. Their depth and generosity have made this book much better. I'm very grateful for their contributions and their friendship.

Hugh Dubberly not only inspired me to delve into systems through his work and publications, but also graciously agreed to write the book's foreword. It is a tremendous honor for me, and I am tremendously grateful.

The Two Waves team has been a dream to work with. In particular, I want to thank Lou Rosenfeld and my editor Marta Justak. (Marta and I only met in person after the manuscript was completed, so much of our collaboration happened in information environments.)

Many people gave me feedback, help, and inspiration at various stages in the development of the book. I'm especially grateful to Chris Baum, Alex Baumgardt, Alan Cooper, Sue Cooper, Abby Covert, Duane Degler, Bill DeRouchey, Andy Fitzgerald, Dave Gray, Jeff Johnson, Jonathan Hoffberg, Priyanka Kakar, Hans Krueger, Andrew Markel, Harry Max, Karen McGrane, Peter Morville, Matt Nish-Lapidus, Brian O'Kelley,

Dan Ramsden, Bob Royce, Samantha Soma, Jeff Sussna, Christina Wodtke, and Richard Saul Wurman.

Finally, I want to thank Jimena, Julia, Ada, and Elias, whose love and support don't falter—even when projects such as this book subtract from our shared lifetimes.

About the Author

 JORGE ARANGO is a strategic designer and information architect. Upon seeing the then-new World Wide Web in 1994, he left his career in (building) architecture to start the first web design consultancy in Central America. He has since designed information environments for organizations that range in scope from developing-world nonprofits to Fortune 500 corporations.

He is co-author (with Louis Rosenfeld and Peter Morville) of *Information Architecture: For the Web and Beyond* (2015), the fourth edition of O'Reilly's celebrated "polar bear" book. He is also a former president of the Information Architecture Institute, and speaks and teaches about design leadership around the world.

Jorge lives with his wife and three children in the San Francisco Bay Area. You can reach him via email at jarango@jarango.com or follow him on Twitter, where he is @jarango.

Colophon

This book was written in various information environments. Given that this is the book's subject, I thought you'd like to know about them.

I used a variety of devices to access a suite of software applications where the development of the book happened. I saved reference materials in Microsoft's OneNote application (https://www.onenote.com), which I could access whether I was on my iPhone, iPad, or Mac. I had a single OneNote notebook for the book, divided into four sections: Notes, Meetings, Admin, and Research. I also used paper notebooks, which I scanned into OneNote in chronological order.

Eastgate's Tinderbox (http://www.eastgate.com/Tinderbox/) application allowed me to think through the structure of the book. Tinderbox enabled me to view my ideas in outline and map views, so I could explore subjects that were linked to each other. If you sensed a structural rhythm and coherence in the book, you can thank this application.

I did the actual writing in Ulysses (https://ulyssesapp.com). This app allowed me to break down the text into smaller chunks and move them around to explore different structural configurations. [It's similar to a better-known app called Scrivener (https://literatureandlatte.com/scrivener/overview).] Ulysses also enabled me to set word targets for each chapter, which gave me small goals to aim for.

I also used various graphics apps to edit and create the book's images; these included OmniGraffle, Sketch, and Apple's Preview app.

This collection of software applications and structures influenced my ability to think about the project. Physically, I was in a small office in the back of my home when I did much of the writing. But I also had access to an information environment that I could enter at will anywhere I had one of my mobile devices. I worked on the book while sitting in physical environments that ranged from BART trains to the cabin of a cruise ship in Alaskan waters. Where *did* I write it, if not in this information environment I built for this purpose?

You'll note that thus far I've described only the information environment where *I* worked. There are other information environments where I shared the work and received feedback.

Ulysses allowed me to easily upload chapter drafts to Medium, where I shared them (privately) with my editor. While useful, I discovered early on that this wasn't a perfect approach. One particular annoyance was that footnotes didn't make it to Medium. Still, this was a very low-friction way of sharing the work I was doing in a format that reads cleanly.

When the book took more form, I shared manuscripts as ebooks. Again, this is something that is easy with Ulysses; the app can export to PDF, ePub, HTML, and Microsoft Word. The latter was important when the time came to share the manuscript with tech reviewers and copyeditors, which we did using Gmail.

Exporting to MS Word marked a major milestone for the project since that meant managing changes in a different system than Ulysses. I often found myself making changes in Word documents and then having to repeat them in the "master" text in Ulysses. As a result, I now have an "imperfect" version of the manuscript.

Even as I write these words, other people are combing through the text to correct last-minute spelling and grammatical errors. I probably won't see these changes until I get a proof of the book. So *Living in Information* has left the nest I created for it. I've gotten used to living in this information environment for the past two years, but now it's time to move on.

www.ingramcontent.com/pod-product-compliance
Lightning Source LLC
Jackson TN
JSHW011927131224
75386JS00033B/1097

* 9 7 8 1 9 3 3 8 2 0 6 5 1 *